Aha! Insight

啊哈，
灵机一动

〔美〕马丁·伽德纳　著
胡作玄　评点

科学出版社
北京

图书在版编目(CIP)数据

啊哈，灵机一动 /（美）伽德纳（Gardner, M.）著；胡作玄评点 . —北京：科学出版社，2006
（20 世纪科普经典特藏）
ISBN 978-7-03-016666-1

I. 啊… II. ①伽…②胡… III. 数学 – 普及读物 IV. 01-49

中国版本图书馆 CIP 数据核字（2007）第 015285 号

责任编辑：胡升华 郝建华 / 责任校对：宋玲玲
责任印制：苏铁锁 / 封面设计：黄华斌

科学出版社出版
北京东黄城根北街 16 号
邮政编码：100717
http://www.sciencep.com
北京凌奇印刷有限责任公司印刷
科学出版社发行 各地新华书店经销
*
2006 年 1 月第 一 版 开本：720×1000 1/16
2022 年 7 月第十一次印刷 印张：16 3/4
字数：332 000
POD定价： 49.00元
（如有印装质量问题，我社负责调换）

序

20世纪在科学发展史上是一个辉煌的世纪，以物理学和生物学的创新性成果为标志的科学成就，极大地改变了世界的面貌，改变了人类的认知水平、生产方式和生活方式。20世纪也是科学史上的一个英雄世纪，一大批别具一格的科学大师风云际会，相继登场，使科学的舞台展现出前所未有的绚丽风采。20世纪发生了两次世界大战，第二次世界大战催生的原子弹，使社会公众了解了科学的巨大威力，也促使人类认真地审视科学，了解到科学必须要与人类的良知，与人文精神结合在一起，只有合理地利用，才能造福于人类，才能有利于和平，有利于人类社会的可持续发展。进入20世纪80年代，人类更进一步认识到必须携起手来保护生态，控制环境污染，探索可持续发展的道路。可持续发展理念的形成，是20世纪阶级社会发展观进步的一个重大的事件。

回顾20世纪科学走过的道路，从突飞猛进的科学创造，到科学与人文伦理的深度撞击，形成与人文精神交融并进的局面，最终在人类文明史上留下了不同寻常的篇章。

20世纪诞生的科学和思想大师所取得的非凡的科学成就、创造的充足科学和思想养分，孕育了一批优秀的科普作品，为公众提供了丰富的精神食粮。人们可以跟着爱因斯坦、薛定谔、伽莫夫、沃森、温伯格、霍金等科学大师的生花妙笔去领略科学创造的历程，登攀一个个科学顶峰的征程和科学高峰的神奇景观；可以跟着卡逊在寂静的春天里思考知更鸟的命运；

可以跟着萨根去观察宇宙和生命……今天这些科学大师和思想大师大部分都已离开了我们，但那些优秀科普作品是他们留给后代的不朽的精神财富。

20 世纪已经过去，21 世纪已经肯定是一个全球化、知识化的世纪，也是一个科技国际化、网络化的时代。可持续发展依然是人类唯一的发展道路，自然科学、社会科学、人文精神将交叉融合，世界的文化环境会发生很大的变化，东西方文化将会在激荡过程中进一步融合升华，创造出具有国际化，又有民族特色的新文化。在未来 15 年，中国要基本完成向一个创新型国家过渡。建立创新体系、创新机制配套的基础是要大幅度提高国民的文化教育水平和科学素质，把我国庞大的人口负担真正转化为无可比拟的创新人力资源。

在中国这样一个大国传播普及科技知识、科学精神是一个宏大的系统工程，需要政府组织倡导和社会各界的积极努力。中国科学院也承担着光荣而艰巨的任务，我们有义务整合全院资源努力把科普工作做大、做好，为国家和社会发挥更大的作用。科学出版社是科普图书出版的一支战略方面军，应该大有作为。《20 世纪科普经典特藏》把原汁原味的经典科普大餐奉献给新时代读者，辅之以中文点评是一个很好的尝试。希望这些经典著作能给读者以启发，开拓读者的科学视野，更希望这些经典著作能起到示范的作用，推进我们自己的原创科普和科学文化作品的创作和出版。

路甬祥

2006年2月十七日

马丁·伽德纳
Martin Gardner

本书作者马丁·伽德纳（Martin Gardner）是当代最著名的数学科普作家之一。他于1914年生于美国俄克拉荷马州的塔尔萨，中学时代就对数学深感兴趣，并一直保持至今。由于他想成为一位物理学家，因此，他没有接受过正式的高等数学教育。他进入芝加哥大学之后，逐渐对科学哲学产生了兴趣，因此放弃了物理学，专攻哲学，并于1936年获得学士学位。毕业后，他从事新闻工作。1941年美国参战，他应征入海军服役4年。战后，他回到芝加哥大学读哲学研究生，但未取得学位。其后8年他主要是自由撰稿人，特别是为儿童杂志"Humpty Dumpty"（这是在英美众所周知的矮胖子的形象）撰稿来维持生计。

1956年，数学界出现两件大事改变了他的一生。当时由纽曼（Newman）主编的四大卷《数学世界》（*The world of Mathematics*）成为英美的畅销书。也正是在这件事的影响下，有着110年历史的著名科普杂志《科学美国人》（*Scientific American*）的主编皮尔（Gerard Piel）看到了数学科普的商机，决定创办《数学游戏》专栏。由于此前伽德纳曾写过一两篇数学方面的文章给《科学美国人》，因此，皮尔邀请他主持这个专栏。后来他回忆说，他根本就没有准备好，当时他连一本有关数学游戏的书都没有。于是，他跑到纽约，买下所有有关书籍。事情就这样开始了，这成为他的终身事业。

他说，他喜欢写这个专栏是因为他热爱数学。但万万没有想到的是，《数学游戏》专栏受到广泛的欢迎，成为《科学美国人》的招牌产品。他本人也出了名，结交了许多大数学家，也受到许多业余数学爱好者的注意。从 1956 年到 1981 年，他几乎每月一篇，连续不断地写了 25 年。1981 年年底他退休后，每年还偶尔写上一两篇。这些文章现在大都收集在一起，形成了十几本单行本。另外，他又写了十几本书，如《数学狂欢节》(*Mathematical Carnival*) 等，而本书 “Aha! Insight” (1978) 是其中最著名的，曾被译成法文、德文、俄文、日文等多种外文，中文译本《啊哈！灵机一动》(白英彩、崔良沂译)于 1981 年由上海科学技术文献出版社出版。除了数学科普之外，他还写了许多一般科普著作，值得注意的是，他与伪科学进行不懈的斗争，特别是那些打着科学名义贩卖伪科学私货的人。

由于他在数学科普方面的贡献，他荣获 1987 年美国数学会斯蒂尔（Steele) 奖和 1994 年数学交流奖。

胡作玄

数学游戏帮你学数学

在漫长的学数学的过程中，写作业已经压得学生喘不过气来，哪还有时间读点伽德纳关于数学游戏的书呢？大多数人不去读，他们在老师、家长的督促下的确可以继续走下去。可是，多数人没有兴趣，没有主动性，更不会问学数学究竟为什么，自己究竟有什么提高？

学过9年或12年数学之后，也许能够解许多练习题，甚至考一个好分数，但是，思想上有什么提高呢？有人说，数学是思想的体操，学过数学之后，思想上应该有个飞跃，可是媒体上不时报道这样的消息，大学生积极参加骗人的传销，也有人相信以前的彩票结果对今后的彩票有影响等等。在日常生活中，更多的人不知如何理财使自己获益最大。这些都显示，数学还要从教科书以外去学习，教科书与数学游戏书是互补的。

伽德纳的书对于读者有什么好处呢？我想首先是激发你的兴趣。很少人会对数学教科书有兴趣，但多数人会对数学游戏感兴趣。20世纪末，匈牙利人的魔方风靡世界就是一个证明。魔方的数学比较复杂，伽德纳的书则十分简单，容易入手。大多数人对下棋、扑克牌有兴趣，而数学游戏则更为简单，而且可以一个人玩。这要比沉溺于电脑游戏更有利于身心。其次，数学游戏激发你的主动性。课本甚至奥校的题目，想出来正确的答案也就完了，但是许多数学游戏可以永远地做下去，只要你还有兴趣。中国发明的幻方就是一种。幻方在中国叫“纵横图”，也就是把1，2，

3，…，n个数排列在$n \times n$个方格中，使得n行，n列以及两条对角线上数码之和都相等。这样得到的结果称为n阶幻方。3阶幻方本质上只有一种，4阶幻方就有880种，因此造出幻方对每个有兴趣的人都是一个挑战。直到最近，仍有许多大数学家研究各种巧妙的构造n阶幻方的方法。因此，只要去做，永远有做不完的事。第三，数学游戏帮助思想方法的提高。许多数学游戏并不是一下子就能想到的，一旦想到之后，对于思想会有很大的促进。在切蛋糕与火柴拼图的游戏中，我们往往只考虑平面的情形，而一旦想到3维空间去，问题就迎刃而解。还有许多问题试来试去解不出，这时就要考虑它是否没有解，当然这是需要证明的。在数学史上这种“眼前无路想回头”的事屡见不鲜，而且都形成划时代的成就，如古希腊三大作图问题、5次方程一般不能根式解、非欧几何等。当然历史上也有少数人顽固不化，死不回头，等待他们的也只能是失败。

有了这么多好处，快来看这本书吧！除了上面的好处之外，还有一个，那就是同时可以学不少英文！

胡作玄

Contents 目录

Introduction 前言

The creative act owes little to logic or reason. In their accounts of the circumstances under which big ideas occurred to them, mathematicians have often mentioned that the inspiration had no relation to the work they happened to be doing. Sometimes it came while they were traveling, shaving or thinking about other matters. The creative process cannot be summoned at will or even cajoled by sacrificial offering. Indeed, it seems to occur most readily when the mind is relaxed and the imagination roaming freely.

Morris Kline, *Scientific American*, March 1955.

引言是全书的核心，在日常生活中，在科学、艺术、商业、政治等活动中更多需要直觉，更多需要灵机一动，而这是从教科书中和数学练习中难以学到的。而这正是本书的用处。它能教会你 aha! 灵机一动。

Experimental psychologists like to tell a story about a professor who investigated the ability of chimpanzees to solve problems. A banana was suspended from the center of the ceiling, at a height that the chimp could not reach by jumping. The room was bare of all objects except several packing crates placed around the room at random. The test was to see whether a lady chimp would think of first stacking the crates in the center of the room, and then of climbing on top of the crates to get the banana.

The chimp sat quietly in a corner, watching the psychologist arrange the crates. She waited patiently until the professor crossed the middle of the room. When he was directly below the fruit, the chimp suddenly jumped on his shoulder, then leaped into the air and grabbed the banana.

The moral of this anecdote is: A problem that seems

difficult may have a simple, unexpected solution. In this case the chimp may have been doing no more than following her instincts or past experience, but the point is that the chimp solved the problem in a direct way that the professor had failed to anticipate.

At the heart of mathematics is a constant search for simpler and simpler ways to prove theorems and solve problems. It is often the case that a first proof of a theorem is a paper of more than fifty pages of dense, technical reasoning. A few years later another mathematician, perhaps less famous, will have a flash of insight that leads to a proof so simple that it can be expressed in just a few lines.

Sudden hunches of this sort—hunches that lead to short, elegant solutions of problems—are now called by psychologists "aha! reactions". They seem to come suddenly out of the blue. There is a famous story about how William Rowan Hamilton, a famous Irish mathematician, invented quaternions while walking across a stone bridge. His aha! insight was a realization that an arithmetic system did not have to obey the commutative law. He was so staggered by this insight that he stopped and carved the basic formulas on the bridge, and it is said that they remain there in the stone to this day.

Exactly what goes on in a creative person's mind when he or she has a valuable hunch? The truth is that nobody knows. It is some kind of mysterious process that no one has so far been able to teach to, or store in, a computer. Computers solve problems by mechanically going step-by-step through a program that tells them exactly what to do. It is only because computers can perform these steps at such incredible speeds that computers can solve certain problems that a human mathematician cannot solve because it might take him or her several thousand years of nonstop calculation.

The sudden hunch, the creative leap of the mind that "sees" in a flash how to solve a problem in a simple way, is something quite different from general intelligence. Recent studies show that persons who possess a high aha! ability are all intelligent to a moderate level, but beyond that level there seems to be no correlation between high intelligence and aha! thinking. A person may have an extremely high I. Q., as measured by standardized tests, yet rate low in aha! ability. On the other hand, people who are not particularly brilliant in other ways may possess great aha! ability. Einstein, for instance, was not particularly skillful in traditional mathematics, and his records in school and college were mediocre. Yet the insights that produced his general theory of relativity were so profound that they completely revolutionized physics.

获得灵感的心态最主要是：先设法找出一个答案，然后看是否能够把方法简化。当动了脑筋之后，也就会使你的解题能力大大提高。

This book is a careful selection of problems that seem difficult, and indeed are difficult if you go about trying to solve them in traditional ways. But if you can free your mind from standard problem solving techniques, you may be receptive to an aha! reaction that leads immediately to a solution. Don't be discouraged if, at first, you have difficulty with these problems. Try your best to solve each one before you read the answer. After a while you will begin to catch the spirit of offbeat, nonlinear thinking, and you may be surprised to find your aha! ability improving. If so, you will discover that this ability is useful in solving many other kinds of problems that you encounter in your daily life. Suppose, for instance, you need to tighten a screw. Is it necessary to go in search of a screwdriver? Will a dime in your pocket do the job just as well?

The puzzles in this collection are great fun to try on friends. In many cases, they will think for a long time about a problem, and finally give it up as too difficult. When you tell them the simple answer, they will usually laugh. Why do they laugh? Psychologists are not sure, but studies of creative thinking suggest some sort of relationship between creative ability and humor. Perhaps there is a connection between hunches and delight in play. The creative problem solver seems to be a type of person who enjoys a puzzling challenge in much the same way that a person enjoys a game of baseball or chess. The spirit of play seems to make him or her more receptive for that flash of insight that solves a problem.

Aha! power is not necessarily correlated with quickness of thought. A slow thinker can enjoy a problem just as much, if not more, than a fast thinker, and he or she may be even better at solving it in an unexpected way. The pleasure in solving a problem by a shortcut method may even motivate one to learn more about traditional solving techniques. This book is intended for any reader, with a sense of humor, capable of understanding the puzzles.

There certainly is a close connection, however, between aha! insights and creativity in science, in the arts, business, politics, or any other human endeavor. The great revolutions in science are almost always the result of unexpected intuitive leaps. After all, what is science if not the posing of difficult puzzles by the universe? Mother Nature does something interesting, and challenges the scientists to figure out how she does it. In many cases the solution is not found by exhaustive trial and error, the way Thomas Edison found the right filament for his electric light, or even by a deduction based on the relevant knowledge. In many cases the solution is a Eureka insight. Indeed, the exclamation "Eureka!" comes from the ancient story of how Archimedes suddenly solved an hydraulic problem while he was taking a bath. According to the legend, he was so overjoyed that he leaped out of the tub and ran naked down the street shouting "Eureka! Eureka!" (I have found it!)

We have classified the puzzles of this book into six categories: combinatorial, geometric, number, logic, procedural and verbal. These are such broad areas that there is a certain amount of unavoidable overlap, and a problem in one category could just as well be regarded as in one of the others. We have tried to surround each puzzle with a pleasant, amusing story line intended to put you in a playful mood. Our hope is that this mood will help you break away from standard problem solving routines. We urge you, each time you consider a new puzzle, to think about it from all angles, no matter how bizarre, before you spend unnecessary time trying to solve it the long way.

After each problem, with its delightful illustrations by the Canadian graphic artist Jim Glen, we have added some notes. These comments discuss related problems, and indicate how, in many cases, the puzzles lead into significant aspects of modem mathematics. In some cases, they introduce problems that are not yet solved.

We have also tried to give some broad guidelines for the channels along which aha! thinking sometimes moves:

1. Can the problem be reduced to a simpler case?

2. Can the problem be transformed to an isomorphic one that is easier to solve?

3. Can you invent a simple algorithm for solving the problem?

4. Can you apply a theorem from another branch of mathematics?

5. Can you check the result with good examples and counterexamples?

6. Are aspects of the problem given that are actually irrelevant for the solution, and whose presence in the story serves to misdirect you?

We are rapidly entering an age in which there will be increasing temptation to solve all mathematical

Gardner 教你 6 个办法去解题。当然这需要灵活运用。

Gardner 的问题共有65组，可以分为6类。但分类不一定很严格，差不多所有题都是综合的。这使你要灵活运用所学过的东西。

Gardner 问题中有些是很简单甚至无聊的问题，如第 6 组的一些问题；也有一些解决了就完了的，缺少变化的问题，如火柴图形变化问题；真正好的问题是能够继续发展下去能激发你更进一步探索的问题，如幻方、几何剖分等。许多数学家都是由此起步得出重要结果的。

problems by writing computer programs. The computer, making an exhaustive trial-and-error search, may solve a problem in just a few seconds, but sometimes it takes a person hours, even days, to write a good program and remove all its bugs. Even the writing of such a program often calls for aha! insights. But with the proper aha! thinking, it may be possible to solve the same problem without writing a program at all.

It would be a sad day if human beings, adjusting to the Computer Revolution, became so intellectually lazy that they lost their power of creative thinking. The central purpose of this collection of puzzles is to exercise and improve your ability in this technique of problem solving.

Chapter 1

Combinatorial aha!

组　　合

Puzzles about arrangements

Combinatorial analysis, or combinatorics, is the study of how things can be arranged. In slightly less general terms, combinatorial analysis embodies the study of the ways in which elements can be grouped into sets subject to various specified rules, and the properties of those groupings.

For example, our first problem is about the ways in which differently colored balls can be grouped together. This problem asks the reader to find the smallest sets of colored balls that have certain properties. The second problem concerns ways in which players can be grouped on a chart for an elimination tournament—a problem with important counterparts in the computer sorting of data.

Combinatorial analysis often asks for the total number of different ways that certain things can be combined according to certain rules. The "enumeration problem", as this is called, is introduced in the episode about the number of ways that Susan can walk to school. In this case, the elements to be combined are the segments of a path along the edges of a matrix. Since geometrical figures are involved, we are in the area of "combinatorial geometry".

Every branch of mathematics has its combinatorial aspects, and you will find combinatorial problems in all the sections of this book. There is combinatorial arithmetic, combinatorial topology, combinatorial logic, combinatorial set theory—even combinatorial linguistics, as we shall see in the section on word play. Combinatorics is particularly important in probability theory where it is essential to enumerate all possible combinations of things before a probability formula can be found. There is a famous collection of probability problems called *Choice and Chance*. The word "choice" in the title refers to the book's combinatorial aspect.

Our very first problem concerns probability because it asks for an arrangement of colored balls that makes

每一章开头的文字一般是十分概括的，事先可以粗看一遍，读完后再仔细领会会更有收获！组合问题有许多，最基本的就是“计数”，也就是每种模式有多少种？这问题看来简单，实际上并不简单，特别是要区分开不同的类型，不要重复计数，也不要漏掉计数，“完全性”是十分重要的。

certain (that is, have a probability equal to 1) a specified task. The text suggests how endless other probability questions arise from such simple questions as the number of ways objects can be put together. Enumerating Susan's paths to school provides a close link to Pascal's triangle and its use in solving elementary probability questions.

The number of arrangements that solve a given combinatorial problem obviously can be none, one, any finite number, or an infinite number. There is no way to combine two odd integers so that their sum is odd. There is only one way to combine two prime numbers so that their product is 21. There are just three ways to combine two positive integers so their sum is 7. (They are the pairs of opposite faces on a die.) And there is an infinite number of combinations of two even numbers that have an even sum.

Very often in combinatorial theory it is extremely difficult to find an "impossibility proof" that no combination will meet what is demanded. For example, it was not until recently that a proof was found that there is no way to combine the planar regions of a map so that the map requires five colors. This had been a famous unsolved problem in combinatorial topology. The impossibility proof required a computer program of great complexity.

On the other hand, many combinatorial problems that seem at first to be difficult to prove impossible can sometimes be proved easily if one has the right aha! insight. In the problem of "The Troublesome Tiles", we see how a simple "parity check" leads at once to a proof of combinatorial impossibility that would be hard to obtain in any other way.

The second problem about the defective pills ties combinatorial thinking into the use of different base systems for arithmetic. We see how numbers themselves and the way in which they are represented in positional notation by numerals depend on combinatorial rules. Indeed, all deductive reasoning, whether in mathematics or pure logic, deals with combinations of symbols in a "string", according to the rules of a system that decides whether the string is a valid or invalid assertion. This is why Gottfried Leibniz, the seventeenth-century father of combinatorics, called the art of reasoning an *ars combinatoria.*

A Sticky Gum Problem

泡泡糖问题

第一个问题并不是个数学问题，而是智力测验，也就是aha，让你思考，灵机一动。
这个问题对读者的启发是随便拿到的2粒糖，会有几种不同的情况？

Poor Mrs. Jones tried to get past the bubble gum machine before her twins noticed it.

First Twin: Mommy, I want some gum.

Second Twin: Me too, Mom. And I want the same color Billy gets.

The penny gum machine is almost empty. There is no way of knowing the color of the next ball. If Mrs. Jones wants to be sure of getting a pair of matching balls, how many pennies *must* she be prepared to spend?

Mrs. Jones could get 2 red balls by spending 6 cents—4 cents to get all the white balls out and 2 cents to get a pair or red. Or she could get 2 white balls by spending 8 cents. So she must be prepared to spend 8 cents, right?

Wrong. If the first 2 balls don't match, the third has to match one of the first 2. So 3 pennies are the most she needs to spend.

Now suppose the machine contains 6 red balls, 4 white, and 5 blue. Can you figure out how many pennies Mrs. Jones needs to have on hand to be sure of getting a pair of matching gum balls?

Did you get 4 cents? If so, you can start thinking about Mrs. Smith who tried to walk by the same gum machine with her triplets.

This time the machine contains 6 red balls, 4 white, and just one blue. How many pennies must Mrs. Smith be ready to spend to get three matching balls?

▥ blue
■ red
□ white

How Many Pennies?

The second gum ball problem is an easy variation of the first one, and is solved by the same insight. In this case the first *three* balls could be of different colors—red, white and blue. This is the "worst" case in the sense that it is the longest sequence of drawings that fail to achieve the desired result. The fourth ball will necessarily match one of the three. Since it could be necessary to buy four balls to get a matching pair, Mrs. Jones must be prepared to spend four cents.

这问题是好的数学游戏问题，这表现在它可以推广到更一般的情形：
① 糖的颜色可以是2种，3种，…，n种。
② 孩子数可以是2人，3人，…，m人。
③ 每种颜色糖各有不同的数目。

The generalization to n sets of balls, each set a different color, is obvious. If there are n sets, one must be prepared to buy $n+1$ balls.

The third problem is more difficult. Instead of twins,

Mrs. Smith has triplets. The gum machine contains 6 red balls, 4 white, and 1 blue. How many pennies might she have to spend to get *three* matching balls?

As before, we first consider the worst case. Mrs. Smith could get 2 red balls, 2 white, and the single blue ball, making 5 in all. The sixth ball must be red or white, so it is sure to make a triplet of the same color, therefore the answer is six cents. Had there been more than one blue ball, she could have drawn a pair of each color, requiring a seventh ball to complete the triplet.

The aha! insight is "seeing" the length of the "worst" case. One might try to solve the problem a harder way by assigning a letter to each of the 11 balls, then examining all possible drawing sequences to see which one has the longest initial chain before a triplet appears. But this method of solving the problem would require listing 11! = 39,916,800 sequences! Even if one approached the problem by not distinguishing between balls of the same color, it would still be necessary to list 2,310 sequences.

The generalization to matching k-tuplets is as follows. If there are n sets of balls (each a different color, and each containing at least k balls), then to obtain a matching k-tuplet one must draw $n(k-1) + 1$ balls. You may enjoy investigating what happens when one or more of the color sets contain less than k balls.

Problems of this sort can be modeled in many other ways. For example: How many cards must you draw from a 52-card deck to be sure that you have, say, 7 cards of matching suit? Here $n = 4$, $k = 7$. The formula gives the answer: $4(7-1) + 1 = 25$.

Although these are simple combinatorial puzzles, they lead into interesting and difficult probability questions. For instance, what is the probability that you will get 7 cards of the same suit if you draw n cards (n ranging from 7 through 24) without replacing each card after it is drawn? (Obviously, the probability is 0 if you draw fewer than 7 cards, and 1 if you draw 25 or more.) How do the probabilities alter if cards are replaced and the deck shuffled after each drawing? A more difficult question: What is the expected number (average in the long run) of drawings you have to make to get k (cards of the same suit), with or without replacement?

The Ping Pong Puzzle

乒乓赛难题

这是一个理论上重要、应用广泛的问题。Knuth 的名著 *The Art of Computer Programming* 有专章论述。本书第一次出现二进制。许多问题与它不可分！

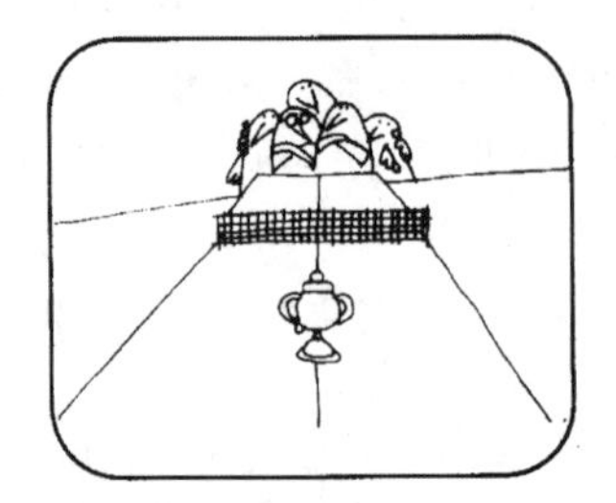

The 5 members of the Millard Fillmore Junior High School ping pong club decided to hold an elimination tournament.

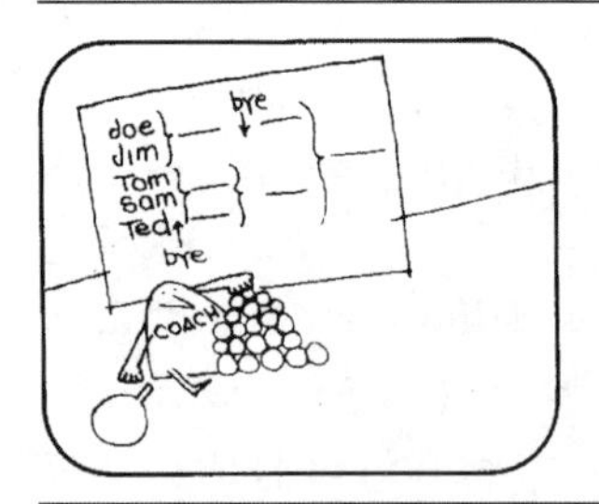

The coach explained his tournament chart this way.

Coach: Five is odd, so one player gets a 'bye' in the first round. And there has to be another 'bye' in the next round. So altogether 4 matches must be played.

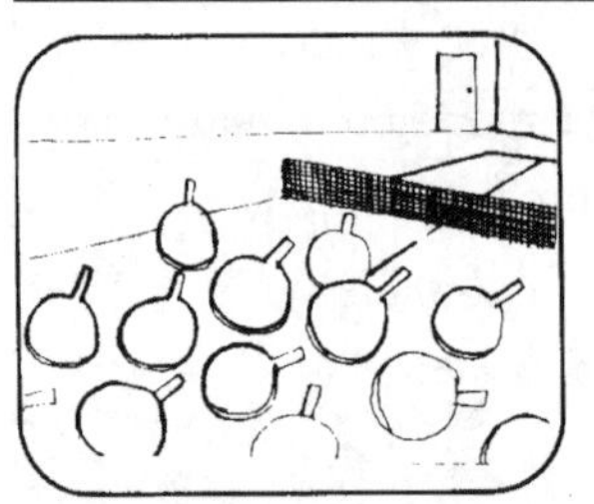

Table tennis was so popular next year that the club had 37 members. Again the coach designed a tournament with the smallest possible number of 'byes'. Can you figure out the number of games that were played?

You don't have it worked out yet? You're still drawing your chart? You've missed an aha! Each match eliminates one player and because there are 36 players to be eliminated, there has to be just 36 games, doesn't there?

How Many Byes?

If you worked on this problem the hard way, by drawing up actual charts of a tournament for 37 players, you may have noticed that no matter how the chart is drawn there are always just 4 byes. The number of necessary byes is a function of *n*, the number of players. How can the number of byes be calculated?

Given *n*, the number of byes can be determined as follows. Subtract *n* from the lowest power of 2 that is equal to or greater than *n*. Express this remainder in binary notation. The number of byes is equal to the number of ones in this binary expression. In our case, we subtract 37 from 64 (which is 2^6) to get 27. In binary notation 27 = 11011. There are four 1's, therefore the tournament must have four byes. It is an interesting exercise to justify this curious algorithm.

The type of tournament described in this problem is often called a "knockout tournament". It corresponds to what computer scientists call an algorithm for determining the largest element in a set of *n* elements by comparing them pairwise. As we have seen, exactly $n-1$ pairwise comparisons are necessary for determining the maximum. Computer sorting can also be done by comparing sets in groups of three, four, five and so on.

The topic of sorting is so important in computer science and its applications that entire books have been written about it. You can easily think of many practical problems in which sorting procedures are important. It is estimated that about one-fourth the running time of computers that are used in science, business and industry is spent on sorting problems.

Quibble's Glasses

奎贝尔的玻璃杯

Barney, who works at a soda fountain, is showing two of his customers a puzzle that uses ten glasses.

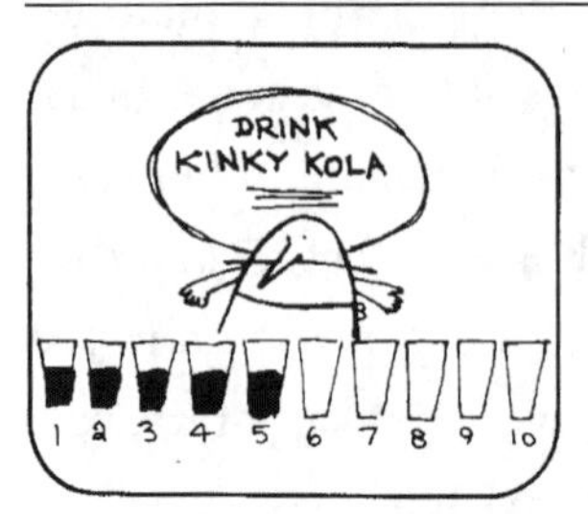

Barney: There are ten glasses in this row, the first five are filled with Kinky Kola, the next five are empty. Can you move just 4 glasses to make a row in which the full and empty glasses alternate?

Barney: That's right. Just switch places with the second and seventh glass, and with the fourth and ninth.

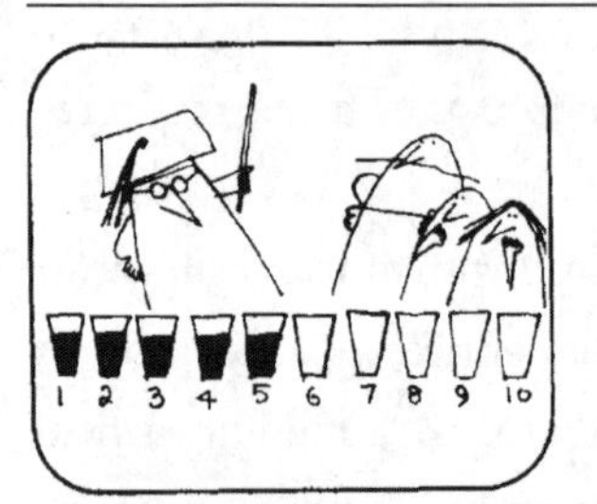

Professor Quibble, who was always thinking of tricky solutions, happened to be listening.

Prof. Quibble: Why four glasses? I can do it by just moving two glasses. Can't you?

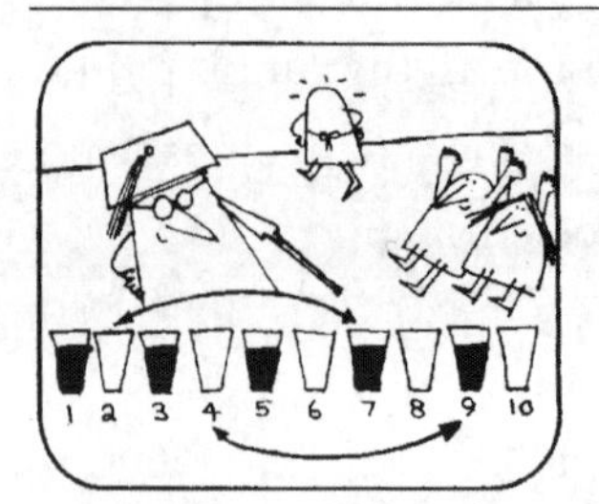

Prof. Quibble: It's simple. Just pick up the second glass and pour its contents into the seventh. And then pick up the fourth and pour into the ninth.

Non-trivial Quibble

Although Professor Quibble solved the puzzle by a verbal quibble, the original problem is not as trivial as it first seems. For example, consider the same problem with 100 full glasses in a row next to 100 empty glasses. How many switches of pairs are necessary to arrange the row so that full and empty glasses alternate?

Quibble 的原意是模棱两可，一语双关。这里他的着数属于出奇制胜之类。

Since it is impractical to work on the problem with 200 glasses, the first step is to analyze the situation for smaller values of n, where n is the number of filled (or empty) glasses, and look for a pattern. You can work on the problem by using counters of two different colors. (Face-up and face-down cards can also be used, or coins that are heads and tails, or two different values.) The problem requires no moves if $n = 1$, and has an obvious solution of one move when $n = 2$. You may be surprised to discover that one switch also solves the problem for $n = 3$. With a little more effort, you may hit on the simple pattern. When n is even, the number of required switches is $n / 2$, and when n is odd, it is $(n-1)/2$. Therefore, if there are 100 full and 100 empty glasses, the problem is solved in 50 switches.

This requires moving 100 glasses. Quibble's joke method of solving the problem cuts the number of required glasses in half.

There is a classic puzzle very similar to the one just analyzed, but harder to solve. Begin with the same row of n objects of one type, adjacent to n objects of another type. (As before this can be modeled with glasses, counters, cards and so on.) You wish to change the row to alternating objects, but now we define "move" differently. In this case, you must slide any adjacent *pair* of counters to any *open* position on the row, without altering the order of the two counters that are moved.

For example, here is how it is done when $n = 3$:

X X X ○ ○ ○
　　X ○ ○ ○ X X
　　X ○ ○　　X ○ X
　　　○ X ○ X ○ X

What is the general solution? It is trivial when n = 1, and you will quickly find that it is not solvable when n = 2. For all n greater than 2, the puzzle is solved in a minimum of n moves.

It is not an easy problem to find a solution when n = 4, and you will enjoy searching for one. Perhaps you can formulate a procedure for solving the puzzle in n moves when n is 3 or more.

如果是真正的置换问题那就是一个好的数学问题，有许多应用。

Many unusual variations of the problem provide other challenges. Here are a few:

1. The rules are the same as before except that when you move each adjacent pair of counters, you switch the positions of the two counters if they are different colors. Thus a black-red pair becomes red-black before you finish the move. With 8 counters there is a solution in five moves. For 10 counters five moves also suffice. We know of no general solution. Perhaps you can find one.

2. The rules are the same as in the original problem except there are n counters of one color and n + 1 counters of the other color, and only pairs of unlike colors may be moved. It has been proved that for any n the puzzle can be solved in n moves, and that this is minimal.

3. Counters of three different colors are used. Pairs of adjacent counters are moved in the usual way to bring all the colors together. If n = 3 (9 counters in all), there is a solution in five moves. In this and all previous variations, we assume that there are no gaps in the final row. If gaps are permitted, there is a surprising solution in four moves.

Other variants suggest themselves which, so far as we know, have not been proposed before, let alone solved. For instance, one could move three or more adjacent counters at a time, and apply this move to any of the above variants.

And what happens if one moves one counter, then two adjacent counters, then three, then four, and so on? Given n counters of one color and n of another, can it always be solved in n moves?

Perplexing Paths
令人困窘的道路

Susan has a problem. When she walks to school she keeps meeting Stinky.

Stinky: Hi Susan. Can I walk with you?

Susan: No. Please go away.

这是一个典型的组合问题，作者的标记法既避免重复计数又避免漏掉，是一个好方法。

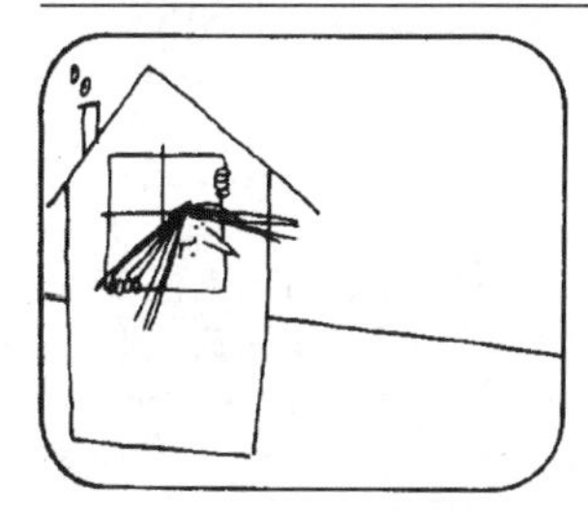

Susan: I know what I'll do. I'll walk to school in a different way every morning. Then Stinky won't know where to find me.

This map shows all the streets between Susan's house and her school. On this particular path Susan is always walking east or south.

Here is Susan on another path. Naturally she doesn't want to walk away from the school. But how many paths are there?

Susan: I wonder how many different ways I can go. Let's see. Hmm. This is going to be tough to figure out. Hmm. Aha! It's not hard at all. It's simple! What insight did Susan have?

Here's how she reasoned.

Susan: I'll put a 1 at the corner where I live because I have just one way to start. Then I'll put a 1 at each corner that's one block away because there's only one way to get there.

Susan: Now I'll put a 2 at this corner because I can get to it in two different ways.

When Susan noticed that 2 is the sum of 1 and 1, she suddenly realized that the number on every corner must be the sum of the one or two nearest numbers along the paths leading to that corner.

Susan: There four more corners are labeled. I'll soon finish the others. Can you complete the labelling of the corners for Susan and tell her how many different ways she can walk to school?

How Many Paths?

上述 Susan 方法的应用实例。

The remaining five vertices, reading top-down and left-right, are labeled 1, 4, 9, 4 and 13. The 13 at the last vertex shows that Susan has 13 ways to walk to school along shortest paths.

What Susan discovered is a simple, fast algorithm for calculating the number of shortest paths from her house to school. Had she attempted to draw all these paths, then count them, it would have been tedious, and out of the question if the street grid had contained a very large number of cells. You will better appreciate the algorithm's efficiency if you actually trace all 13 paths.

To test your understanding of the algorithm, try sketching a variety of other street networks and applying the algorithm to determine the number of shortest paths from any vertex A to any other vertex B. Figure 1 gives four problems of this type. They can be solved in other ways, using combinatorial formulas, but the methods are tricky and complicated.

1

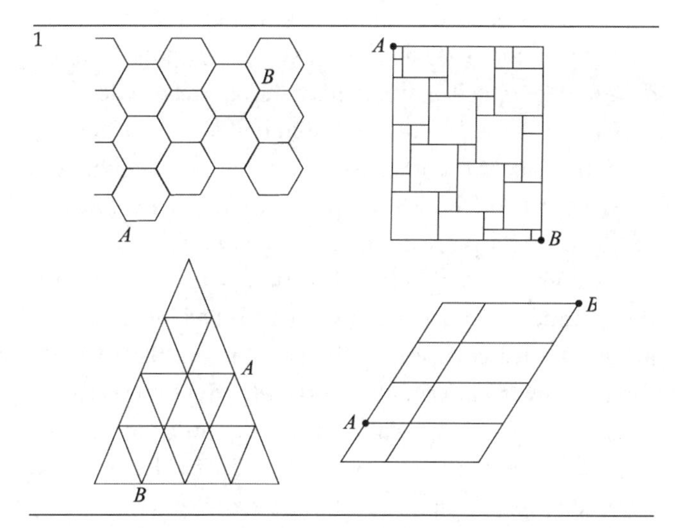

What is the number of the shortest paths by which a chess rook can move from one corner of a chessboard to the diagonally opposite corner? This problem is quickly solved by labeling all the cells of the board in the same manner that Susan labeled the street corners. A chess rook moves only along orthogonals (horizontally and vertically), therefore the shortest paths are obtained by confining each move to a direction that carries the rook toward its goal. When the entire board has been labeled correctly, as shown in Figure 2, the labels will give at once the number of the shortest paths from the starting

2

1	8	36	120	330	792	1716	3432
1	7	28	84	210	462	924	1716
1	6	21	56	126	252	462	792
1	5	15	35	70	126	210	330
1	4	10	20	35	56	84	120
1	3	6	10	15	21	28	36
1	2	3	4	5	6	7	8
♖	1	1	1	1	1	1	1

square to any square on the board. The cell at the upper right corner has the number 3,432, therefore there are 3,432 ways that the rook can go from one corner to the diagonally opposite corner along the shortest routes.

Let us slice the chessboard in half along a diagonal, then turn it so it becomes the triangle shown in Figure 3. The numbers on the bottom row of cells give the number of the shortest paths from the apex cell to each cell at the bottom. The labeling of this triangle is identical with the numbers of Pascal's famous number triangle. The algorithm for computing the shortest paths from the top downward is, of course, precisely the procedure by which Pascal's triangle is constructed. This isomorphism provides an excellent introduction to the endless fascinating properties of the Pascal triangle.

最后显示组合学的基础二项式系数及杨辉三角形。

3

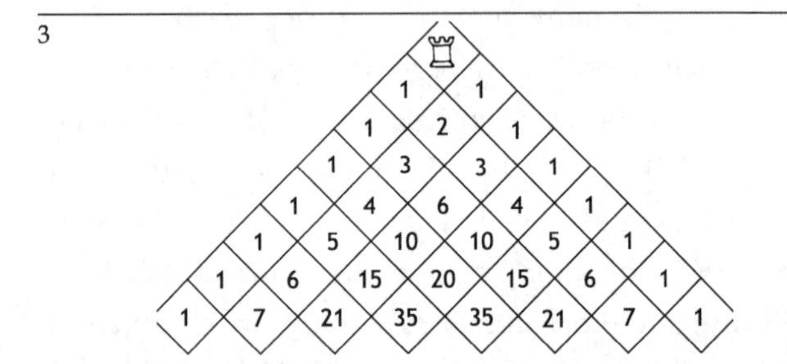

Pascal's triangle gives at once the coefficients for the expansion of binomials—that is, raising $(a + b)$ to any power—as well as the solutions to many problems in elementary probability theory. Note that in Figure 3 the number of the shortest paths from the top of the triangle to the bottom row of cells is 1 on the outside border cells, and the numbers increase as you move toward the center. Perhaps you have seen one of those devices based on Pascal's triangle in which a board is tipped and hundreds of little balls roll past pegs to enter columns at the bottom. The balls arrange themselves in a bell-shaped binomial distribution curve precisely because the number of the shortest paths to each slot are the coefficients of a binomial expansion.

Susan's algorithm obviously works just as well on three-dimensional grids with cells that are rectangular parallelepids. Imagine a cube that is 3 units on the side, and divided into 27 unit cubes. Consider this a chessboard with a rook in one of the corner cells. The rook can move parallel to any of the three coordinates. In how many ways can it take a shortest path to the cell that is opposite it along a space diagonal?

The Bewildered Babies
搞错了的婴儿

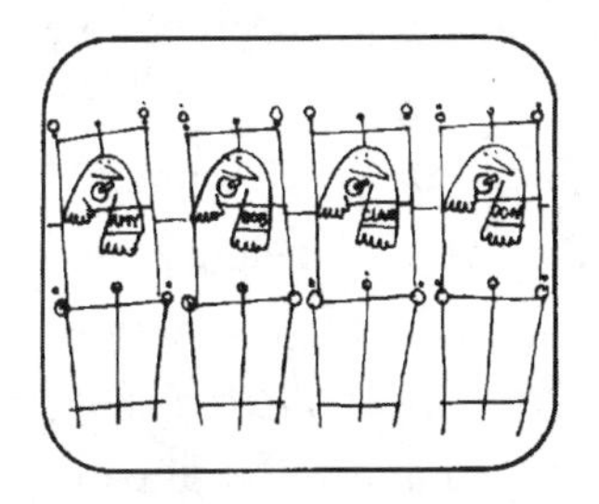

In a certain hospital the identification tags of 4 babies got mixed up. Two babies were tagged correctly and the other two wrong. In how many different ways can this happen?

这也是典型的组合问题。

An easy way to figure this out is to make a chart listing all of the possibilities. It turns out that there are only 6 different ways of tagging just 2 babies incorrectly.

Now suppose that after the tags are mixed up, exactly three are correct and only one wrong. In how many different ways can this happen?

Did you draw up a chart to work this out? Or did you discover the aha!?

Mixed-Up Tags

The reason why this problem confuses so many people is that they assume wrongly that there are many ways that just three out of four babies can be correctly tagged. But if you think in terms of the "pigeon-hole principle", the answer is obvious. Suppose there are four pigeon holes, and each labeled with the name of one of four objects. If three objects are placed in their proper holes, then the fourth object has only one spot it can go, and of course that is its correct spot. Instead of many possible cases, there is only one: Namely the case in which all four objects are correctly placed.

There is a classic mislabeling puzzle, involving only three objects, that is also solved by an aha! insight which reduces the number of cases to one. Suppose you have three closed boxes on a table. One contains two nickels, one contains two dimes, and one contains a nickel and a dime. The boxes are labeled 10¢, 15¢ and 20¢. But every label is *incorrect*. Someone reaches into the box mislabeled 15¢, removes one coin, and places it on the table in front of the box. By seeing this coin, can you tell the contents of each box?

As before, one is at first inclined to think there are many different possibilities, but with the right insight you can see that there is only one case. The coin taken from the mislabeled 15¢ box must be either a nickel or a dime. If it is a nickel, you know that the box originally held two nickels. If it is a dime, you know the box originally held two dimes. In either case, the contents of the other two boxes are now fully determined. To understand why, draw up a chart of the six possible cases. You will see that the mislabeling of all three boxes eliminates all but two cases. The sampling of one coin from the 15¢ box then eliminates an additional case, leaving only the correct one.

This problem is sometimes given in a slightly more difficult form. One is asked to determine the contents of all three boxes by sampling a minimum number of coins which may be taken from any box. The unique answer, of course, is one coin taken from the 15¢ box. Perhaps you can invent more complicated versions with more than two objects per box, or more than three boxes.

Many other fascinating puzzles are closely related to the baby problem, and that also lead into elementary probability theory. For example: If the tags of the babies are mixed at random, what is the probability that all four will be correct? That all will be incorrect? That at least one will be correct? That exactly one will be correct? That at least two will be correct? That exactly two will be correct? That at most two will be correct? And so on.

The "at least one" question, in general form, is one of the classics of recreational mathematics. It is often given with a story about n men who check their hats at a restaurant. A careless hat-check girl makes no attempt to match hats with checks, but hands out the checks randomly. What is the probability that at least one man gets his own hat back? It turns out that the probability quickly approaches a limit of $1-(1/e)$ as n increases, or a little better than $1/2$. Here, e is a famous irrational constant, called Euler's constant, equal to 2.71828^{+}. It is as frequently encountered in probability problems as *pi* is in geometrical problems.

Quibble's Cups

奎贝尔的塑料杯

Professor Quibble has a puzzle for you.

Prof. Quibble: Take 3 empty styrofoam coffee cups and try to put 11 pennies in them so that each cup holds an odd number of pennies.

又是 Quibble! 他想到在一个杯子中还套上另外的杯子。

Prof. Quibble: That wasn't so hard, was it? There are lots of ways to do it. You could put 3 pennies in one cup, 7 in another, and 1 in the third.

Prof. Quibble: However, can you put 10 coins in the same cups so that there is an odd number in each cup? It is possible, but you'll have to think in a tricky way to do it.

Prof. Quibble: I hope you didn't give up. All you had to do was to think of putting 1 cup inside another. Now, isn't it easy to arrange the cups so that there's an odd number in each cup?

Quibble's Subset

The aha! that solves Quibble's brain teaser is the realization that by putting one cup inside another, the same set of coins can belong to more than one cup. In the language of set theory, our solution is a set of 7 elements plus another set of 3 elements that contains a subset of 1 element. This solution can be represented with circles as follows:

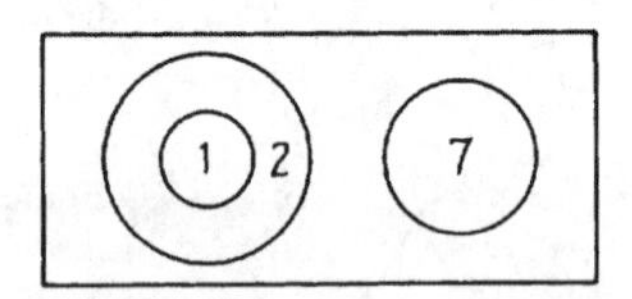

You will enjoy finding all the other solutions. It is easy to find 10 of them, of which the above is one, but it will take another aha! to discover that there are five more, or 15 in all.

After you have found all 15 you might try generalizing the puzzle by varying the number of coins, the number of cups, and the rule about the kinds of numbers to go in each cup.

The basic insight—that part or all of one set can be included in another set and counted twice—is involved in many famous puzzles and paradoxes. Here is an amusing one.

这一问题显示重复计数所造成的悖论。

After a boy failed to attend school for several weeks he was visited by the school's attendance officer. The boy explained why he had no time for school.

"I sleep 8 hours a day. That makes 8×365 or 2,920 hours. There are 24 hours per day, so that's the same as 2,920/24 or about 122 days."

"Saturday and Sundays are not school days. That amounts to 104 days per year."

"We have 60 days of summer vacation."

"I need 3 hours a day for meals—that's 3×365 or 1,095 hours per year, or 1,095/24 which is about 45 days per year."

"And I need at least 2 hours per day for recreation. That comes to 2×365 or 730 hours, or 730/24 which is about 30 days per year."

The boy jotted down these figures and added up all the days:

Sleep	122
Weekends	104
Summer	60
Meals	45
Recreation	30
	361

The total came to 361 days.

"You see," said the boy, "that leaves me only 4 days to be sick, and I haven't even considered the school holidays we get every year!"

The attendance officer studied the boy's figures but couldn't find anything wrong with them. Try this paradox on your friends to see how many of them can spot the fallacy: Namely, counting subsets more than once. The boy's categories overlap like the contents of Quibble's cups.

Steak Strategy

炙肉片策略

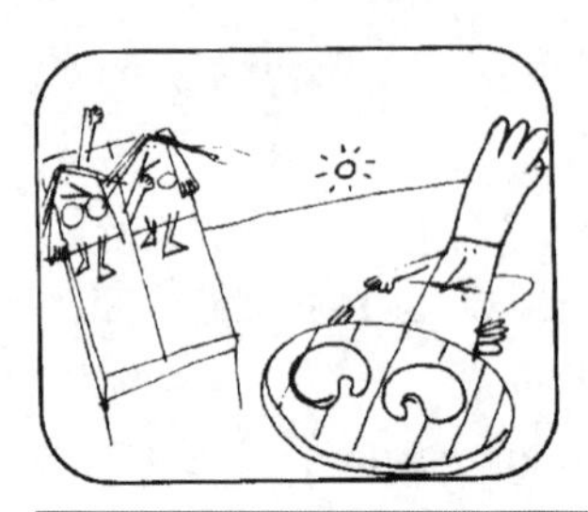

Mr. Johnson has a small outdoor grill just big enough to hold two steaks. His wife, and daughter Betsy, are hungry and anxious to eat. The problem is to broil 3 steaks in the shortest possible time.

Mr. Johnson: Let's see. It takes 20 minutes to broil both sides of one steak because each side takes 10 minutes. And since I can cook two steaks at the same time, 20 minutes will be enough time to get two steaks ready. Another 20 minutes will broil the third steak and the job will be done in 40 minutes.

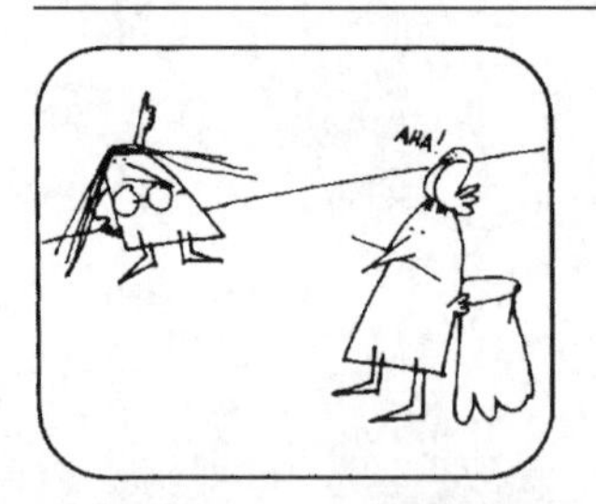

Betsy: But you can do it much faster, Daddy. I just figured out how you can save 10 minutes. What clever aha! insight did Betsy have?

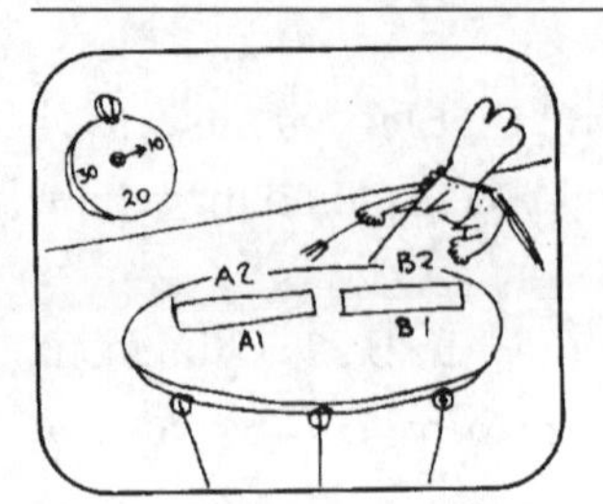

To explain Betsy's solution, we call the steaks A, B, and C; each steak having sides 1 and 2. In the first 10 minutes sides A1 and B1 are broiled.

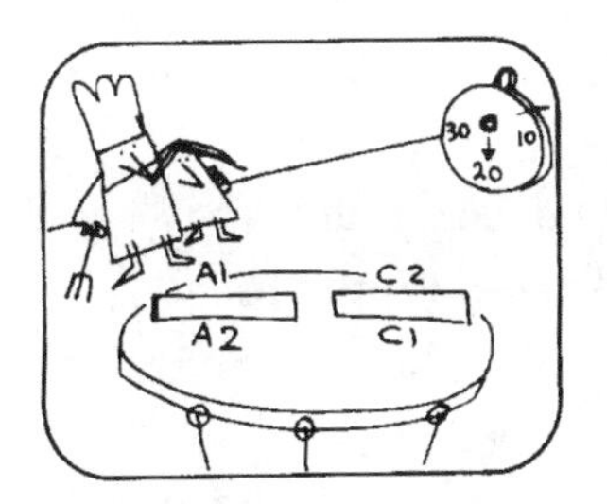

Steak B is now put aside. And in the next 10 minutes sides A2 and C1 are broiled. Steak A is now finished.

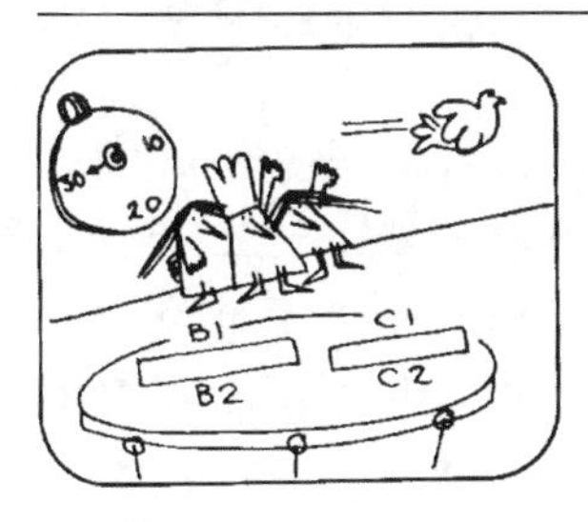

Ten more minutes and sides B2 and C2 are broiled. All three steaks are cooked in only 30 minutes, right?

A General Strategy

This is a simple combinatorial problem in a branch of modern mathematics called operations research. It brings out beautifully the fact that if one is faced with a series of operations, and wants to complete them in the shortest time, the best way to schedule the operations is not immediately apparent. The way that seems at first to be the best, may be considerably improved. In our problem the aha! lies in recognizing that it is not necessary to cook the second side of a steak immediately after cooking the first side.

严格讲这不是组合问题而是排序问题，是运筹学的主要问题。运筹学处理的主要是优化问题，即在一定条件下，使目标函数达到最大或最小。许多问题经过推广是典型的数学问题，而且还有尚未解决的。

As usual, simple problems like this can be generalized in more ways than one. For example, you can vary the number of steaks the grill will hold, or vary the number of steaks to be cooked, or both. Another generalization is to consider objects with more than two sides, and which have to be "finished" in some way on all sides. For example, a person may have the task of painting n cubes red on all side, but at each step he can paint only the tops of k cubes.

Operations research is used today for solving practical problems in business, industry, military strategy, and many other fields. To appreciate the usefulness of even a solution as simple as the one for our steak problem, consider the following variation.

Mr. and Mrs. Jones have three household tasks to perform.

1. Their first floor must be vacuumed. They have only one vacuum, and the task takes 30 minutes.

2. The lawn must be mowed. They have only one mower, and this task also takes 30 minutes.

3. Their baby must be fed and put to bed. This, too, requires 30 minutes.

How should they go about performing these tasks so as to accomplish all of them in a minimum amount of time? Do you see how this problem is isomorphic with the steak problem? Assuming that Mr. and Mrs. Jones work simultaneously, one might at first suppose it would require 60 minutes to complete the tasks. But if one task, say vacuuming, is split in half, and the second half postponed (as in the steak problem), the three tasks can be completed in three-fourths the time, or 5 minutes.

Here is a more sophisticated operations research problem involving the preparation of three slices of hot buttered toast. The toaster is the old-fashioned type, with hinged doors on each of its two sides. It holds two pieces of bread at once but toasts each of them on one side only. To toast both sides it is necessary to open the doors and reverse the slices.

It takes 3 seconds to put a slice of bread into the toaster, 3 seconds to take it out, and 3 seconds to reverse a slice without removing it. Both hands are required for each of these operations, which means it is not possible to put in, take out, or turn two slices simultaneously. Nor is it possible to butter one slice while another is being put into the toaster, turned or taken out. The toasting time for one side of a piece of bread is 30 seconds. It takes 12 seconds to butter a slice.

Each slice is buttered on one side only. No side may be buttered until it has been toasted. A slice toasted and buttered on one side may be returned to the toaster for toasting on its other side. The toaster is warmed up at the start. In how short a time can three slices of bread be toasted on both sides and buttered?

It is not too difficult to figure out a procedure that will do the job in two minutes. However, the total time can be reduced to 114 seconds if you have the following insight: A piece of bread can be partially toasted on one side, removed, and later returned to complete the toasting on the same side. Even with this necessary aha! the task of scheduling the operations in the most efficient way is far from easy. Innumerable practical problems in scheduling are much more complicated than this, and call for very sophisticated mathematical techniques involving computers and modern graph theory.

The Troublesome Tiles

难铺的瓷砖

Mr. Brown's patio is made from 40 square tiles. The tiles have deteriorated and he wants to cover them with a new set.

这个问题第一次向读者显示，并不是任何合理的数学问题都是可解的。

He chooses new tiles to match his lawn furniture. Unfortunately these tiles come only in rectangles, each of which covers two of his old tiles.

Storekeeper: How many of these do you want, Mr. Brown?

Mr. Brown: Well, I have to cover 40 squares. So I'll need 20, I guess.

When Mr. Brown tried to cover his patio with the new tiles he became very frustrated. No matter how hard he tried, he couldn't make them fit.

Betsy: What's the trouble, Dad?
Mr. Brown: These blasted tiles won't fit. It's driving me nuts. I always end up with two squares. I can't cover.

Mr. Brown's daughter drew a plan of the patio and colored it like a checkerboard. Then she studied it for several minutes.

Betsy: Aha! I see what the trouble is. It's obvious once you realize that each rectangular tile must cover a red and a white square.
How does this help? Do you know what Betsy means?

There are 19 black squares and 21 red. So after 19 tiles are placed there will always be two red squares uncovered. And these cannot be covered by the rectangular tile. The only solution is to cut one tile in half.

在许多问题经过多次尝试而得不到解决时，也许可以考虑它是否无解。但是，一般来说，证明不可能性往往是十分困难的。

The Parity Check

Mr. Brown's daughter solved the tiling problem by applying what is called a "parity check". If two numbers are both odd or both even, they are said to have the same parity. If one is odd and the other even they are said to have opposite parity. In combinatorial geometry one often encounters analogous situations.

In this problem two squares of the same color have the same parity, and two squares of opposite color have opposite parity. A rectangular tile clearly covers only pairs of opposite parity. The girl first showed that if 19 rectangular tiles were placed on the patio, the two remaining squares could be covered by the last tile only if the squares were of opposite parity. Since the two remaining squares necessarily have the same parity, they cannot be covered by the last tile, therefore the tiling of the patio is impossible.

Many famous impossibility proofs in mathematics rest on parity checks. Perhaps you are familiar with Euclid's famous proof that the square root of 2 cannot be rational. The proof is obtained by first assuming that the root can be expressed as a rational fraction reduced to the lowest terms. The numerator and denominator cannot both be even because then the fraction would not be in the lowest terms. Therefore they must both be odd, or one must be odd and the other even. Euclid's proof then shows that the fraction cannot be either. In other words, the numerator and denominator cannot be of like parity or of opposite parity. Because every rational fraction must be one or the other, the square root of 2 cannot be rational.

Tiling theory abounds with problems that would be difficult to prove impossible if parity checks were not used. This problem is extremely simple because it involves tiling with dominoes, the simplest non-trivial polyomino. (A polyomino is a set of unit squares attached at their edges.) The girl's impossibility proof applies to any matrix of unit squares which, after a checkerboard coloring, has at least one more cell of one color than it has cells of the other color.

In our problem the patio may be regarded as a 6 by 7 matrix with two missing cells of the same color. Obviously if the two removed cells are the same color, the remaining 40 cells cannot be covered with 20 dominoes. An interesting related problem is whether 20 dominoes will always tile the 6 by 7 matrix if the two removed cells are of *opposite* color. The parity check fails to prove impossibility, but this does not mean that the tiling is always possible. It is out of the question to investigate every possible pattern created by removing a pair of tiles of opposite color, because there are too many possibilities to analyze. Is there a simple proof of possibility for all cases?

Yes, it is both simple and elegant, and one that resulted from a brilliant aha! that occurred to Ralph Gomory. It, too, makes use of a parity principle. Assume that the 6 by 7 rectangle has a closed path, one cell wide, that completely fills it; see Figure 4. Now imagine two cells of opposite color removed from anywhere along the path. This breaks the closed path into two parts. Each part consists of an even number of cells which alternate colors. Clearly such a portion of the path can always be tiled with dominoes. (Think of them as little boxcars that can be arranged along a twisted track.) Therefore the problem is always solvable. You may wish to experiment

4

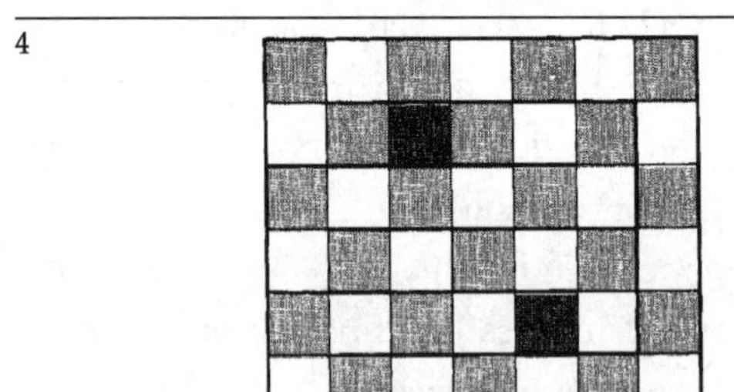

with applications of this clever proof to matrices of other sizes and shapes, and with more than two missing cells.

Tiling theory is a vast area of combinatorial geometry about which there is growing interest. Regions to be tiled can be of any shape—finite or infinite. Tiles may likewise vary in shape, and problems may involve sets of different tiles rather than congruent shapes. Impossibility proofs often involve coloring the field in a specified way with two or more colors.

The three-dimensional analog of a domino is a brick with unit dimensions of $1 \times 2 \times 4$. It is easy to "pack" (tile in space) a $4 \times 4 \times 4$ box with such bricks, but is it possible to pack completely a $6 \times 6 \times 6$ box with such bricks? This problem is answered in exactly the same way as Mr. Brown's patio problem. Imagine the cube divided into 27 smaller cubes, each $2 \times 2 \times 2$. Color these order-2 cubes alternately black and white like a three dimensional checkerboard. If you count the number of unit cubes of each color you will find that there are 8 more cubes of one color than of another.

No matter how a brick is placed within this colored cube, it must always "cover" exactly the same number of black unit cubes as white. But there are 8 more unit cubes of one color than another. No matter how the first 26 bricks are placed, there will always be 8 unit cubes left over of the same color. Therefore they cannot be covered by the 27th brick. This would be extraordinarily difficult to prove by exhaustively examining every possible pattern of packing.

Brick packing theory is only a portion of the theory of tiling in 3-dimensional space. There is a growing literature on space-packing problems, with many tantalizing unsolved questions. Many of the problems have applications to the packing of merchandise in cartons, storage of merchandise in warehouses, and so on.

Parity also plays an important role in particle physics. In 1957 two Chinese American physicists received the Nobel Prize for work that led to the overthrow of a famous law called the "conservation of parity". This is too technical to go into here, but there is a delightful coin trick to illustrate one way parity is conserved.

Toss a handful of coins on the table and count the number of heads. If even, we say the heads have even parity. If odd, we say they have odd parity. Now turn a pair of coins over, then another pair, then another, choosing the pairs at random. You may be surprised to find that no matter how many pairs are reversed, the parity of heads is always conserved. If odd at the start it remains odd. If even at the start it remains even.

This is the basis of a clever magic trick. Turn your back and have someone reverse the coins by pairs for as long as the person likes, then tell the person to cover any coin with a hand. You turn around, and after glancing at the coins you can tell correctly whether the coin under the hand is heads or tails. The secret is to count the number of heads at the outset and remember if the number is odd or even. Since turning coins by pairs does not affect this parity, you have only to count the heads at the finish to know whether the concealed coin is heads or tails.

As a variation, let the person cover two coins with a hand. You can then tell whether the concealed coins are alike or different. Many ingenious card tricks of the mindreading variety operate by similar parity checks.

Quibble's Pets

奎贝尔的动物

Here's Professor Quibble again.

Prof. Quibble: I've got another teaser for you. How many pets do I have if all of them are dogs except two, all are cats except two, and all are parrots except two?

Have you got it?

Professor Quibble has just 3 pets: a dog, a cat, and a parrot. All are dogs except two, all are cats except two and all are parrots except two.

"All" for One

This confusing little problem can be solved in your head if you have the insight that the word "all" can apply to only one animal. The simplest case—one dog, one cat, one parrot—provides the solution. However, it is a good exercise to put the problem into algebraic form.

Let x, y and z stand, respectively, for the number of dogs, cats and parrots, and n for the total number of animals. We can then write four simultaneous equations:

$$n = x + 2$$
$$n = y + 2$$
$$n = z + 2$$
$$n = x + y + z$$

These equations can be solved by any of many standard techniques. It is clear from the first three equations that $x = y = z$. Since $n = x + 2$, and (from the fourth equation) $n = 3x$, we can write $x + 2 = 3x$ which gives x a value of 1. The complete answer follows from this value of x.

Since numbers of animals are usually given in positive integers (who has for a pet a fraction of a cat?) we can think of Quibble's pet problem as a trivial example of what is called a Diophantine problem. This is an algebraic problem with equations that must be solved in integers. Sometimes a Diophantine equation has no solution, sometimes just one, sometimes a finite number greater than one, sometimes an infinite number. Here is a slightly more difficult Diophantine problem that also concerns simultaneous equations and animals of three different kinds.

A cow costs $10, a pig $3 and a sheep 50¢.

A farmer buys 100 animals and at least one animal of each kind, spending a total of $ 100. How many of each did he buy?

Let x be the number of cows, y the number of pigs, z the number of sheep. We can write two equations:

$$10x + 3y + z/2 = 100$$

$$x + y + z = 100$$

Eliminate the fraction in the first equation by multiplying all terms by 2. From this result subtract the second equation. This eliminates x and gives us:

$$19x + 5y = 100$$

What integral values may x and y have? One way to solve this is to arrange the equation with the smallest coefficient on the left: $5y = 100 - 19x$. Dividing both sides by 5 gives: $y = (100 - 19x)/5$. Now divide 100 and $19x$ by 5, putting the remainders (if any) over 5 to form a terminal fraction. The result is:

$$y = 20 - 3x - 4x/5.$$

Clearly the expression $4x/5$ must be integral, and this means that x must be a multiple of 5. The lowest multiple is 5 itself, which gives y a value of 1 and (going back to either of the two original equations) z a value of 94. If x is any larger multiple of 5, y becomes negative. Thus the problem has only one solution: 5 cows, 1 pig and 94 sheep.

You can discover a lot about elementary Diophantine analysis merely by varying the costs of the animals in this problem. Suppose, for instance, cows are \$4, pigs \$2 and sheep a third of a dollar? What animals can the farmer buy for \$100 assuming he buys 100 animals and at least one of each? In this case there are just three solutions. What if cows cost \$5, pigs \$2 and sheep 50¢? Now there is no solution.

Diophantine analysis is an enormous branch of number theory, with endless practical applications. One famous Diophantine problem, known as Fermat's last theorem, asks if there are integral solutions to the equation $x^n + y^n = z^n$ where n is a positive integer greater than 2. (If $n = 2$, it is called a Pythagorean triple, and there are an infinite number of solutions starting with $3^2 + 4^2 = 5^2$.) It is the most famous unsolved problem in number theory. No one has found a solution, or proved that there is not one.

The Medicine Mix-up Small

药品小混

A drugstore received a shipment of ten bottles of a certain drug. Each bottle contains one thousand pills. The pharmacist, Mr. White had just put the bottles on a shelf when a telegram arrived.

Mr. White read the telegram to Miss Black, the store manager.

Mr. White: Urgent. Do not sell any pills until all bottles are checked. By mistake, the pills in one bottle are each 10 milligrams too much. Return the faulty bottle immediately.

Mr. White was annoyed.

Mr. White: Of all the luck, I'll have to take a pill from each bottle and find its mass. What a nuisance!

Mr. White started to do this when Miss Black stopped him.

Miss Black: Wait a minute. There's no need to use the scale 10 times, we only need to use it once.

How is this possible?

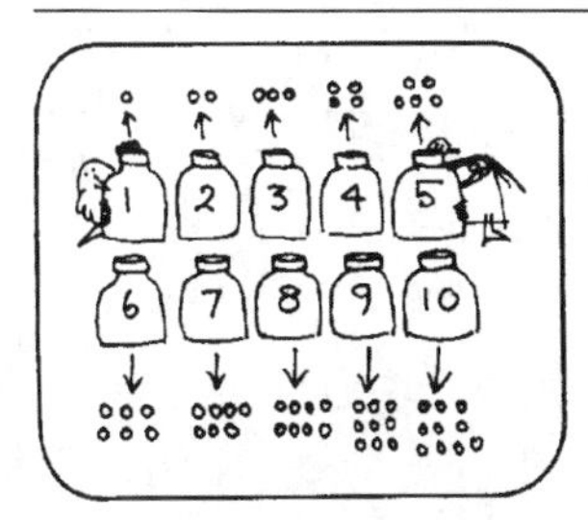

Miss Black's insight was to take 1 pill from the first bottle, 2 from the second, 3 from the third, and so on to 10 from the last bottle.

These 55 pills were put on the scale to find their mass. If it was 5510 milligrams, or 10 milligrams too much, she knew that one pill was too heavy. It had to come from the first bottle.

If the mass was 20 milligrams too much, then 2 pills were too heavy. They had to come from the second bottle. And so on, for the other bottles. So, Miss Black did only use the scale once, didn't she?

The Medicine Mix-Up Big

药品大混

Six months later the store received 10 more bottles of the same pill. Then another telegram arrived saying that a worse mistake had been made.

This time the order included any number of bottles that were filled with pills that were 10 milligrams too heavy. Mr. White was furious.

Mr. White: What shall I do now Miss Black? The system we used before won't work.

Miss Black thought about the problem before replying.

Miss Black: You're right. But if we change the method we can still use the scale only once and identify every faulty bottle. What did Miss Black have in mind this time?

The Drug Dilemmas

In the first pill weighing problem we are told that only one bottle contains heavier pills. By taking a different number of pills from each bottle (the simplest way to do this is to use the sequence of counting numbers) we have a one-to-one correspondence between the set of counting numbers and the set of bottles.

To solve the second problem we must use a sequence that assigns a different number to each bottle, and in addition, every subset of the sequence must have a unique sum. Are there such sequences? Yes, and the simplest is the doubling sequence: 1, 2, 4, 8, 16, ... These numbers are the Successive powers of 2, and the sequence provides the basis of binary notation.

The solution, in this case, is to line the bottles up in a row, then take 1 pill from the first bottle, 2 pills from the second, 4 from the third, and so on. The removed pills are then weighed altogether. Let's assume they are heavier by 270 milligrams. Since each faulty pill is heavier by 10 milligrams, we divide this by 10 to get 27, the number of heavier pills. Write 27 as a binary number: 11011. The position of the 1's tell us what powers of 2 are in the set that sums to 11011 = 27. They are 1, 2, 8 and 16. The 1's are in the first, second, fourth and fifth positions from the right of 11011. Therefore the faulty bottles are bottles l, 2, 4 and 5.

The fact that every positive integer is the sum of a unique set of powers of 2 is what makes the binary notation so useful. It is indispensible in computer science, and in a thousand other areas of applied mathematics.

In recreational mathematics there also are endless applications.

Here is a simple card trick to mystify your friends. Although it may seem to have no connection with the pill bottle problem, the underlying binary principle is the same.

Have someone shuffle a deck of cards. Put the deck in your pocket, then ask anyone to call out a number from 1 through 15. You reach into your pocket and take out a set of cards with values that sum to the number called.

The secret is simple. Before showing the trick, put an ace, deuce, four and eight in your pocket. The deck will be missing four cards, but this is such a small number that their absence will not be noticed. The shuffled deck goes into your pocket beneath the four cards already there. When the number is called, mentally express it as a sum of powers of 2. Thus if 10 is called you think "8 + 2 = 10". Reach into your pocket and take out the deuce and the eight.

Mindreading cards are also based on the same binary principle. Figure 1 of Chapter 3, Number aha! shows a set of six cards that determine any selected number from 1 through 63. Ask someone to think of a number within this range—their age, for instance—then hand you all the cards that bear the number. You immediately name the number. The secret is simply to add the powers of 2 that appear as the first number on each card. For example, if you are handed cards C and F, you sum their two starting numbers, 4 and 32. This tells you that the chosen number is 36.

What rule determines the set of numbers for each card? Every number whose binary representation has 1 in the first position on the right goes on card A, the card whose set of numbers start with 1. These are all the odd numbers from 1 through 63. Card B contains all numbers 1 through 63 whose binary notation has a 1 in the second position from the right. Card C contains all numbers whose binary notation has a 1 in the third position from the right, and so on for cards D, E and F. Note that 63, which is 111111 in binary notation, has a 1 in every position, therefore it appears on every card.

Magicians sometimes make this trick more mysterious by having each card a different color. The magician memorizes the color that stands for each power of 2. For example, the red card is 1, the orange card is 2, the yellow card is 4, the green card is 8, the blue card is 16, and the purple card is 32. (The colors are in rainbow order.) Now the magician can stand across a large room and ask a person to put aside each card on which the thought-of number appears. By noting the colors on the cards placed aside, the magician can immediately call out the chosen number.

The Broken Bracelet

断 金 链

Gloria, a young lady from Arkansas, is visiting in California. She wants to rent a hotel room for a week.

The clerk was very unpleasant.

Clerk: The room is $20 per day and you have to pay cash.

Gloria: I'm sorry sir, but I don't have any cash. However, I do have this solid gold bracelet.

Each of its seven links is worth more than $20.

Clerk: Alright, give me the bracelet.

Gloria: No, not now. I'll have a jeweler cut the bracelet so I can give you 1 link a day. Then when I get some money at the end of the week I'll redeem it.

The clerk finally agreed. But now Gloria had to decide how to cut the bracelet. She was in a dilemma.

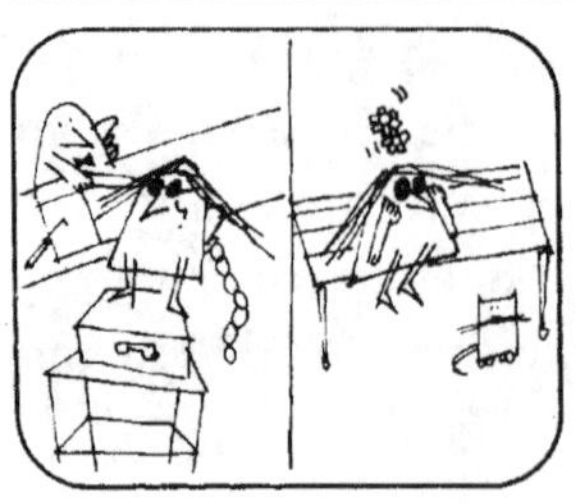

Gloria: I have to be careful because the jeweler is going to charge me for each link that he cuts and for each link that he joins when the bracelet is put back together again.

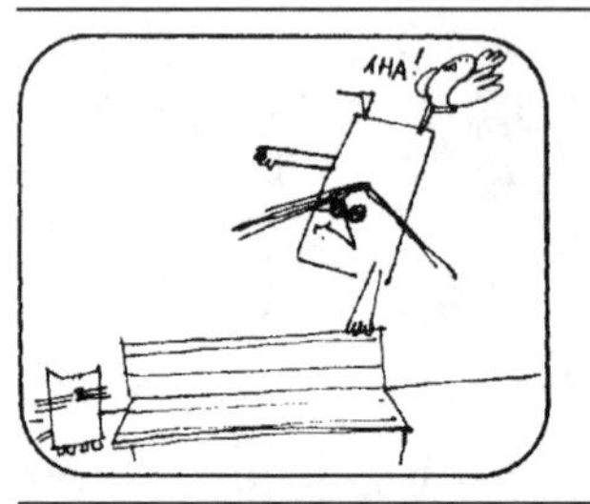

After thinking a while Gloria realized that she didn't have to cut all the links because she could trade pieces back and forth. She couldn't believe it when she figured out how many cuts the jeweler had to make. How many cuts would you make?

Only one link need be cut. It must be third from one end. This makes three pieces of 1, 2, and 4 links. And these are sufficient to trade back and forth so that each day the clerk gets one more link.

Crucial Link

Two aha! insights are needed to solve this problem. The first is to realize that the smallest set of chains that can be combined in various ways to form sets of 1, 2, 3, 4, 5, 6 and 7 links is a set of chains with 1, 2 and 4 links; that is, with numbers in the doubling series. As we learned in the last problem, this is the power series that is the basis of binary notation.

The second insight is to realize that cutting only one link divides the bracelet into this desired set of three chains.

The problem generalizes to chains of longer length. For instance, suppose Gloria had a chain of 63 gold links that she wanted to cut and use in the same way that she used her bracelet—to pay for 63 days, one link per day. The cutting of as few as three links will do the trick. Do you see how? Can you devise a general procedure that solves the problem, with a minimum number of cut links, for a chain of any length?

An interesting variant of the problem is to start with n links that are joined at the ends to make a closed loop. For example, suppose Gloria had a necklace in the form of a closed chain of 79 gold links. How few links need to be cut to pay for 79 days, one link per day?

Chapter 2

Geometry aha!

几　　何

Puzzles about shapes

Geometry is the study of shapes. Although true, this definition is so broad that it is almost meaningless. The judge of a beauty contest is, in a sense, a geometrician because he is judging female shapes, but this is not quite what we want the word to mean. It has been said that a curved line is the most beautiful distance between two points. Even though this statement is about curves, a proper element of geometry, the assertion seems more to be in the domain of aesthetics rather than mathematics.

人们常说，数学的对象是数与形。比起数来，形更为直观。研究形的几何学内容也远为广泛。很长时期内，几何学实际上代表整个数学，这从欧几里得《几何原本》中七、八、九三篇为数论即可看出。 19世纪末，几何学在变换的不变性及群的概念下统一是几何学最重要的成就。

Let us be more precise and define geometry in terms of symmetry. By symmetry is meant any transformation of a figure that leaves the figure unchanged. For example, the letter "H" possesses 180 degree rotation symmetry. This means that if we rotate the letter 180 degrees—turn it upside down—we still have the letter "H". The word "AHA" possesses reflection symmetry. Hold it up to a mirror and the reflection of this word looks the same.

Every branch of geometry can be defined as the study of properties that are unaltered when a specified figure is given specified symmetry transformations. Euclidian plane geometry, for instance, concerns the study of properties that are "invariant" when a figure is moved about on the plane, rotated, mirror reflected, or uniformly expanded and contracted. Affine geometry studies properties that are invariant when a figure is "stretched" in a certain way. Projective geometry studies properties invariant under projection. Topology deals with properties that remain unchanged even when a figure is radically distorted in a manner similar to the deformation of a figure made of rubber.

Although geometry pervades every portion of this book, in this chapter we have brought together problems in which the geometrical aspect dominates, and, of course, we have selected problems that depend on aha! insights for easy solutions. Our first puzzle, about cheese cutting, illustrates how many branches of mathematics can come together in even the simplest problem. It is partly plane geometry, partly solid geometry, partly combinatorial, and partly arithmetical. Moreover, it introduces an important branch of algebra called the "calculus of finite differences".

The "Big Knight Switch," surprisingly, is a problem of topology. The string solution shows that the problem is equivalent to one that can be given in terms of the points on any simple closed curve, and it does not matter in the least what shape the closed curve has. Only the topological properties of such a curve are involved. We solve this problem with points on a circle, but we could just as well have used a square or a triangle.

The next two problems—"Surprising Sword" and "Payoff at the Poles"— take us off the plane once more into three-dimensional Euclidian geometry. The pilot's paths suggest a famous path problem about four bugs that shows how it is sometimes possible to avoid calculus by applying much simpler insights. Ransom's surveying problems take us back to the plane, introducing aspects of Euclidian geometry that belong to dissection theory and tiling theory. The tiling problem is one of combinatorial plane geometry. Miss Euclid's cubeslicing problem is one of combinatorial solid geometry.

The carpet problem, and its three-dimensional companion about the hole in a sphere, are two elegant examples of theorems in which a variable, which one expects to behave like a variable, turns out to have only one value even when other parameters are varied. Who would expect the sphere's volume to be a constant regardless of the hole's width or the radius of the sphere? When a mathematician first encounters this theorem he/ she almost always expresses amazement, followed by the exclamation "Beautiful!"

No one knows exactly what a mathematician means when he/she calls something beautiful—it is somehow bound up with unanticipated simplicity but all mathematicians recognize a beautiful theorem, or a beautiful proof of a theorem, as easily as one recognizes a beautiful person. Geometry, because of its visual aspect, is unusually rich in beautiful theorems and proofs. You will find some good examples of them in this section.

Crafty Cheese Cuts

巧分乳酪

几何学中剖分问题是最重要的一个。2000年有专门的著作问世。

The food at Joe's Diner may not be the best, but the place is famous for its delicious cheese.

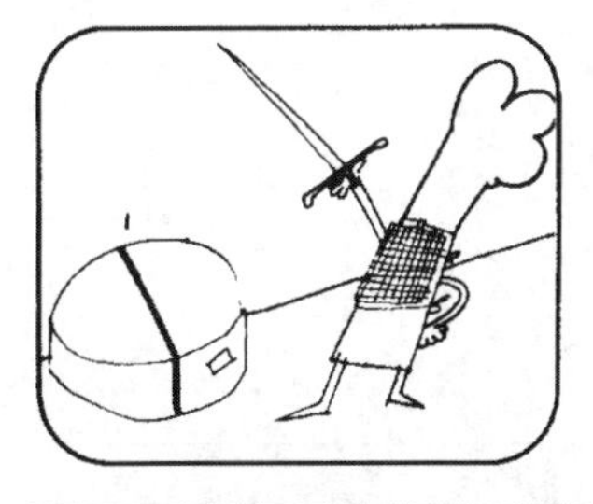

You can have a lot of fun with the cylindrical pieces of cheese. With one straight cut it's easy to divide one piece into two identical pieces.

With two straight cuts it's easy to cut it into four identical pieces. And three cuts will make six identical pieces.

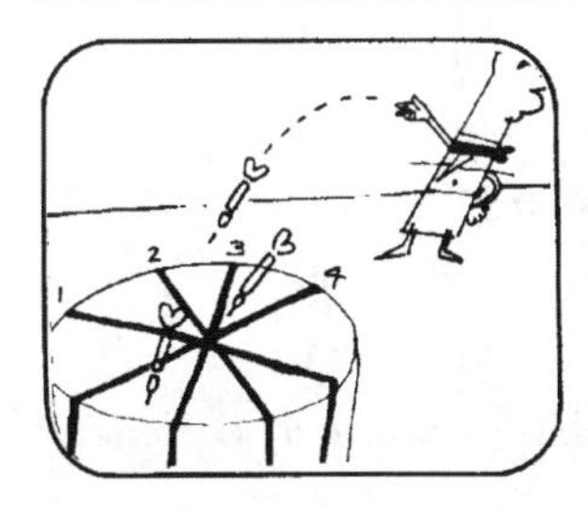

One day, Rosie, the waitress, asked Joe to slice the cheese into eight identical pieces.

Joe: Okay, Rosie. That's simple enough. I can do it with four straight cuts like this.

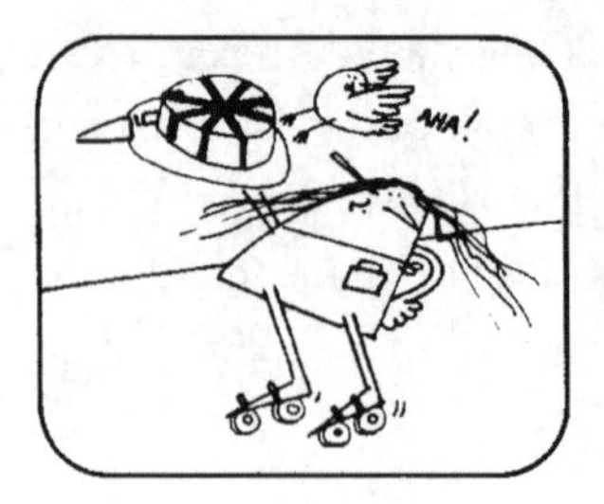

While Rosie was carrying the slices to the table, she suddenly realized that Joe could have gotten the eight identical pieces with only three straight cuts. What insight did Rosie have?

Three Straight Cuts?

Rosie's insight was to realize that the cylindrical cheese is a *solid* figure that can be cut in half by a horizontal plane through the center. Figure 1 shows how three planar cuts divide the cheese into eight identical portions. This solution assumes that the three cuts are made simultaneously. If the cuts are consecutive, and one is allowed to rearrange pieces between cuts, then it can be done in three cuts by stacking the first two pieces, cutting to get four, then stacking the four pieces and cutting to get eight.

1

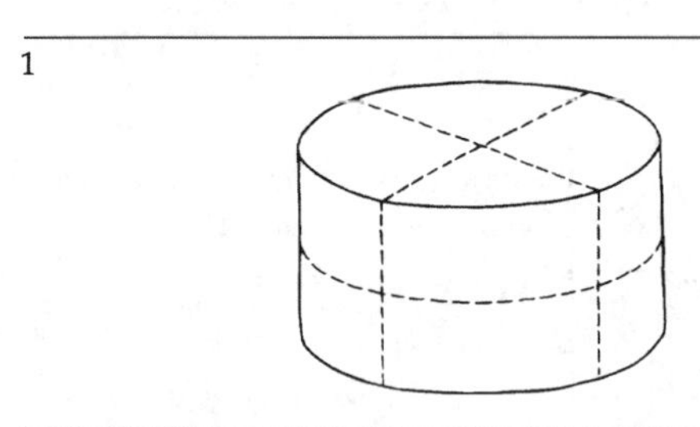

Rosie's solution is so simple that it is almost trivial, yet it provides a good introduction to significant cutting problems that can be explored with the calculus of finite differences and proved by mathematical induction. The calculus of finite differences is a powerful tool for discovering formulas for the general term of number sequences. Today there is a rapidly growing interest in number sequences because of their many practical applications, and because computers can carry out operations with sequences so quickly.

Rosie's first method of slicing the cheese was with straight cuts that are concurrent at the center of the top of the cheese. The top of the cheese is a flat surface like a pancake. Let us see what kinds of number sequences can be generated by the simple procedure of cutting a pancake with straight lines. If the lines are concurrent at the pancake's center, it is obvious that n simultaneous straight cuts produce a maximum of $2n$ pieces.

Does this expression $2n$ also give the maximum number of pieces that can be produced by n concurrent cuts through any plane figure bounded by a simple closed curve? No—as Figure 2 shows, if it's easy to draw nonconvex shapes on which even *one* cut can produce as many pieces as you like. Is it possible to draw a shape such that a single cut will produce any finite number of *congruent* pieces? If so, what characteristics must a shape's perimeter have to permit the formation of n congruent pieces by one straight cut?

2

Cutting a pancake becomes more interesting when the cuts are not concurrent. You will quickly discover that not until $n = 3$ does this procedure start producing more than $2n$ pieces. We are not here concerned with whether the pieces are congruent or even of equal area. Figure 3 shows how the maximum number of pieces is obtained when $n = 1, 2, 3$ and 4. The number of pieces are, respectively, 2 , 4 , 7 and 11.

This is a familiar sequence that is generated by the formula

$$\frac{n(n+1)}{2}+1$$

where n is the number of straight cuts. The first ten terms of the sequence, starting with $n = 0$, are: 1, 2, 4, 7, 11, 16,

22, 29 , 37, 46 , ... Note that the first row of differences is 1,2, 3, 4, 5, 6, 7, 8, 9, ..., and the second row of differences is 1, 1, 1, 1, 1, 1, 1, 1, ... This strongly suggests that the general term for the sequence is a quadratic.

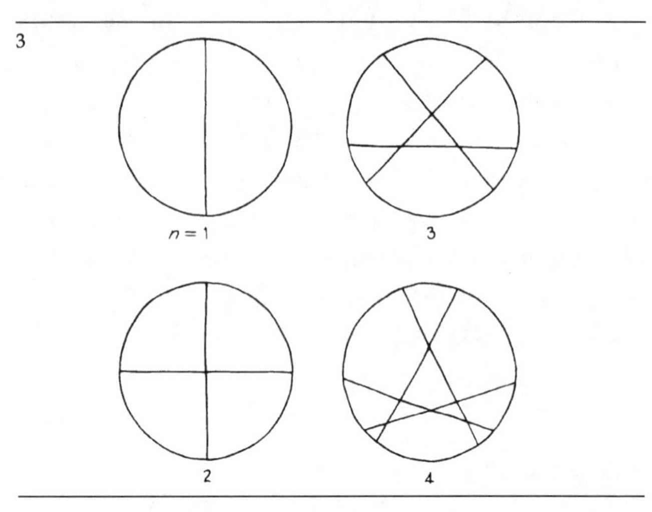

We say "strongly suggests" because finding a formula with the calculus of finite differences does not guarantee that the formula is valid for the infinite sequence. For this a proof is required. In the case of the pancake formula, there is a simple proof by induction that is not hard to work out.

From this point you can take off in dozens of fascinating exploratory directions, many of which lead to unusual number sequences, formulas, and proofs by mathematical induction. Here are a few problems to start on. What is the maximum number of pieces that can be obtained when:

剖分问题随奶酪的形状不同，而有不同的结果，这使它成为一个好的数学游戏。

1. A pancake shaped like a horsehoe is given n straight cuts?

2. A sphere, or cylindrical, piece of cheese like the one cut by Rosie, is given n planar cuts?

3. A pancake is given n cuts with a circular cookie cutter?

4. A pancake shaped like a ring (that is, with a circular hole in the center) is given n straight cuts?

5. A doughnut (torus) is given n planar cuts?

In all these problems it is assumed that the cuts are simultaneous. How do the answers vary if the cuts are consecutive, with rearrangements permitted between cuts?

Dimensions in Disguise

隐蔽的尺寸

In the middle of a city park there is a large circular play area. The city council would like to put a diamond shaped wading pool inside the circular area.

这是平面几何学一个极好的练习，最好引入中学的教材。

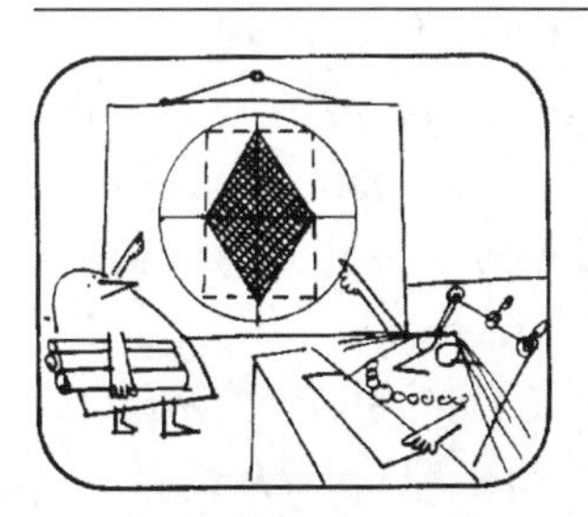

When Doris Wright, the mayor, saw the plans she spoke to the architect.

Mayor Wright: I like the pool's rhombical shape, and the red tiling; but how long is each side of the pool?

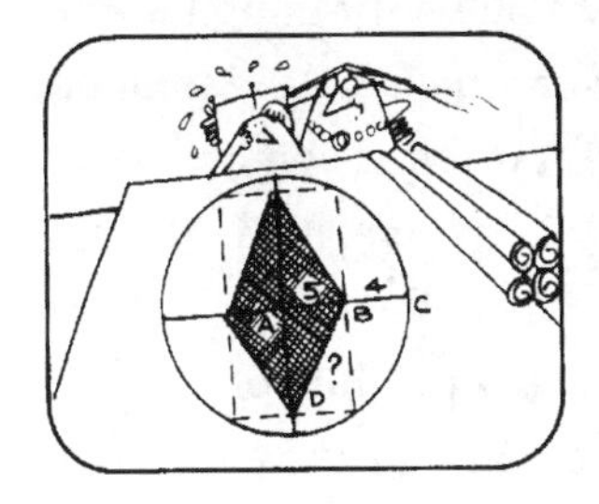

Frank Lloyd Wrong, the architect, was puzzled.

Mr. Wrong: Let's see. It's 5 meters from *A* to *B*, and 4 meters from *B* to *C*. Hmm. There has to be a way to find *BD*. Maybe I have to use Pythagorean theorem.

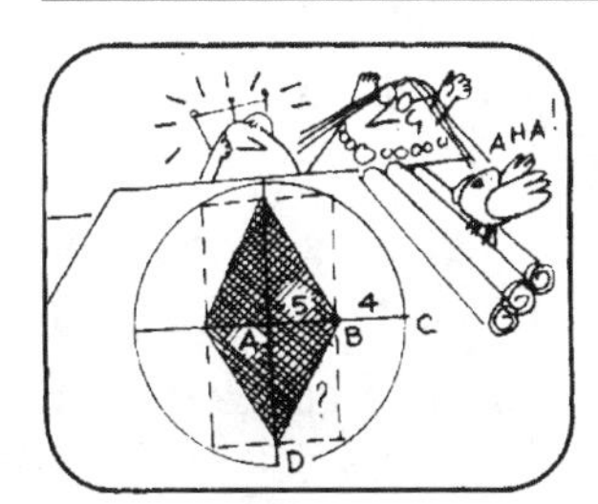

Mr. Wrong was about to give up when suddenly her Honour shouted:

Mayor Wright: Aha! The pool's side is exactly 9 meters.
It's obvious.

Mr. Wrong: By golly, you're Wright. And I'm Wrong.

What simple aha! enabled them to solve the problem so easily?

A Diagonal Radius

Ms. Wright suddenly realized that each side of the pool is the diagonal of a rectangle, and the other diagonal in each rectangle is the radius of the circular play area. The diagonals of a rectangle are equal, therefore the side of the pool is the same length as the circle's radius. The radius is 5 + 4 = 9 meters, therefore each side of the pool is 9 meters. There is no need to apply the Pythagorean theorem.

You can better appreciate the value of this aha! insight by trying to calculate the side of the pool in a more conventional manner. If you use nothing but the Pythagorean theorem and similar triangles, the solution is long and tedious. It can be shortened somewhat by remembering a theorem in plane geometry that says if two chords intersect within a circle, the product of the two parts of one chord equals the product of the two parts of the other chord. This theorem gives the height of the right triangle as $\sqrt{56}$. By applying the Pythagorean theorem you can then calculate the hypotenuse of the right triangle as 9 meters.

A closely related problem is a famous puzzle about a water lily that the poet Henry Longfellow introduced into his novel *Kavenaugh*. When the stem of the water lily is vertical, the blossom is 10 centimeters above the surface of a lake. If you pull the lily to one side, keeping the stem straight, the blossom touches the water at a spot 21 centimeters from where the stem formerly cut the surface. How deep is the water?

The problem can be solved by first drawing the diagram shown in Figure 4. This is essentially the same diagram as the one for the swimming pool problem. Our task is to determine the length of x. Like the pool problem, this also can be solved in more than one way. But if you remember the theorem about intersecting chords, you can solve it with very little effort.

4

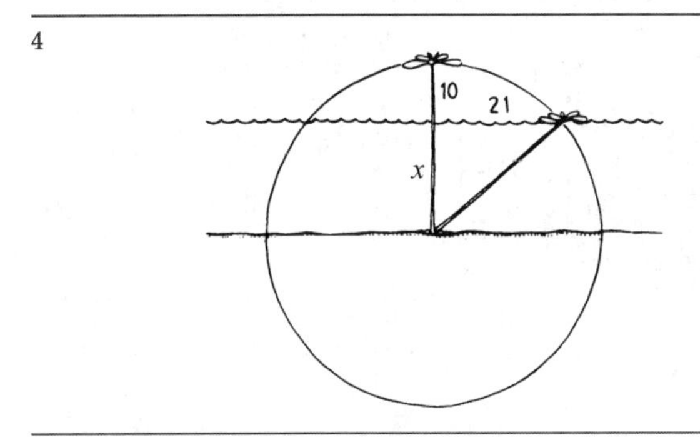

Here is another delightful swimming pool puzzle that is quickly solved by an aha! A dolphin is at the west edge of a circular pool at spot *A*. He swims in a straight line for 12 meters. This causes him to bump his nose against the pool's edge at spot *B*. He turns and swims in a different direction in a straight line for 5 meters, and arrives at spot *C* on the pool's edge exactly opposite *A*, where he first started. How far would he have gone had he swum directly from *A* to *C*?

The aha! that solves the problem is knowing the theorem that an angle inscribed in a semicircle is a right angle, therefore *ABC* is a right triangle. In this case, the sides of the right triangle are given as 5 and 12, therefore the hypotenuse is 13 meters. The moral of all these problems is: In many cases the insight that makes a geometrical problem ridiculously easy depends on remembering a fundamental theorem of Euclidean geometry.

The Big Knight Switch

骑士大调动

国际象棋的马（骑士Knight）也如中国象棋的马一样走“日”字，只是国际象棋的马在格内走，而且没有别马腿的问题。这引出一系列数学游戏问题。

At a meeting of the chess club, Mr. Bishop posed a puzzle.

Mr. Bishop: Interchange the positions of the black and white knights in as few moves as possible.

One boy made his first two moves this way and it took him 24 moves to get the white knights on top and the black knights on the bottom.

Another boy was able to do it in 20 moves.

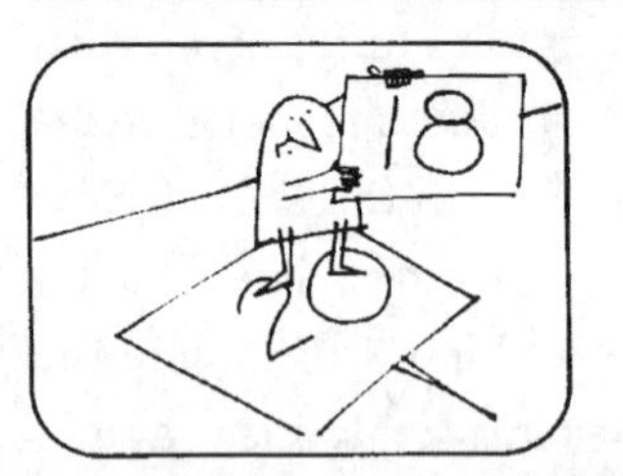

But no one could do it in less than 18 moves until Fanny Fish arrived.

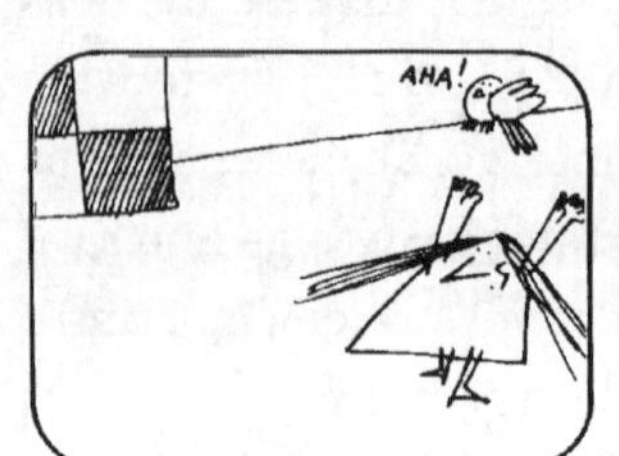

Ms. Fish: Aha! I can do it in only 16 moves. And I can prove that it can't be done in less.

Before starting her explanation Fanny drew a diagram in which straight lines showed every possible knight move.

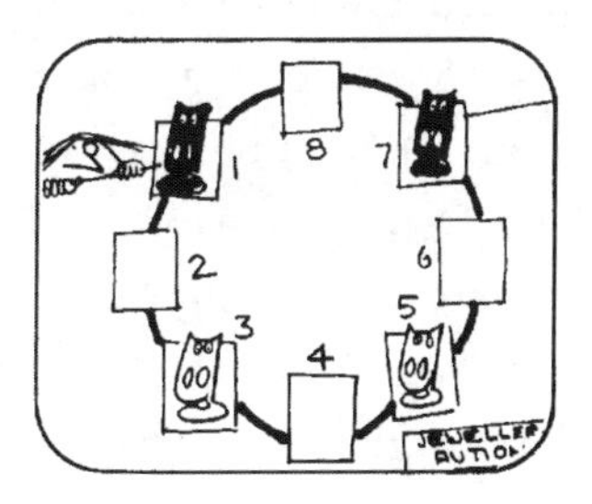

Ms. Fish: If the straight lines are imagined to be strings, the eight cells will be like beads on a folded necklace, which can be opened up to form a circle.

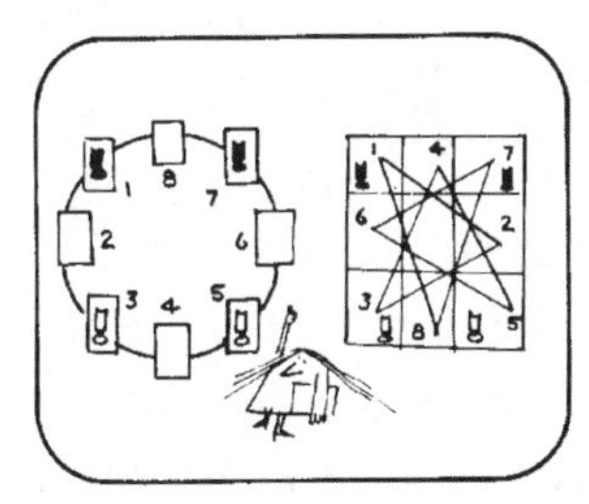

Ms. Fish: Every move on the board corresponds to a move on the circle. To switch the knights we just have to move them around the circle in one direction.

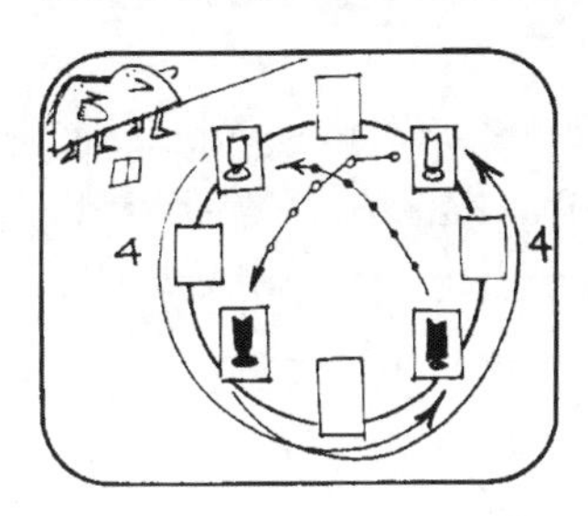

Mr. Bishop: You're right Fanny. And at the finish each of the 4 knights has moved 4 times. That's 16 moves in all and it can't be done in less.

Fanny replaced one of the white knights with a red one, and asked the members to interchange the places of the red and white knights in as few moves as possible. Why do you think she had a smile on her face when she did this?

Knights and a Star

Fanny solved the knights problem by changing it to an isomorphic problem that had a simple aha! solution. The problem she posed is solved by the same curious technique as before. When we join the cells by string, and open them up into a circle, we see that the knights are in the following cyclic order: black, black, red, white. Fanny was smiling because she saw the red and white knights could not interchange their places. Their order is invariant because no knight can jump another knight by moving around the circle in either direction. Do you see why?

这问题表示化为同构问题的重要性。

Going clockwise around the circle, the white knight is always immediately behind the red knight. If it were possible for the red and white knights to exchange starting places, then the cyclic order would have to be reversed and the red knight would be immediately behind the white knight. This obviously is impossible because it would require that one knight hop over both black knights. By changing the problem to one of topological order of four spots on a closed curve, we have found a simple impossibility proof that would be extremely difficult to obtain by any other method. You will surely agree if you try to solve the problem in a different way.

Did you like these two knight switching problems? Here is one that is even more of a challenge. Consider the problem shown on the 3-by-4 board in Figure 5. As before,

5

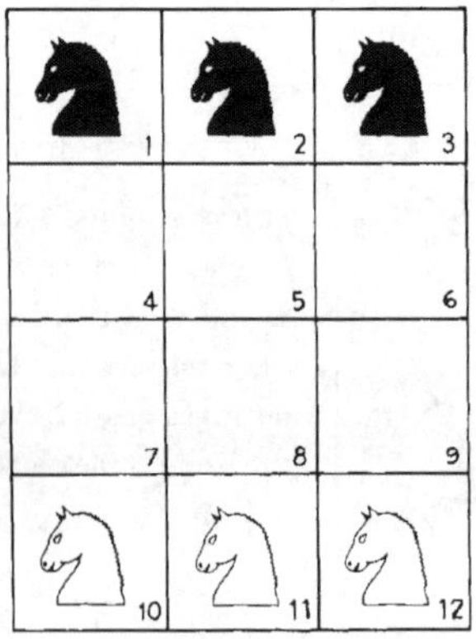

the task is to switch the positions of the three black and three white knights, so that the white knights occupy the top row, and the black knights the bottom row, and to do this in a minimum number of moves.

In this case the isomorphic graph is more complicated, see Figure 6. The graph is, of course, a diagram that

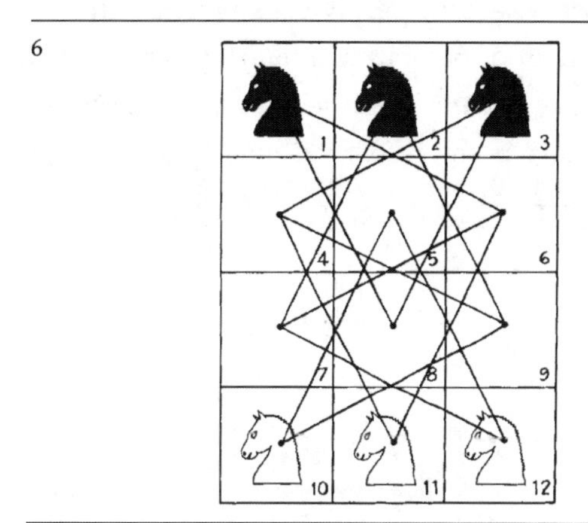

shows every possible knight move. Assuming that the graph is made of strings and beads, we cannot open the string into a circle as we did in the previous problem. However, we can open the bead-and-string graph to the form shown in Figure 7. The numbers in this picture correspond to cell numbers in Figures 5 and 6.

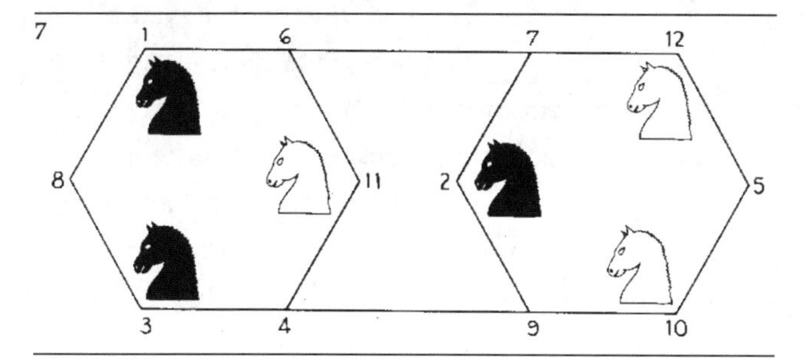

The problem of switching black and white knights on this graph is, therefore, isomorphic with the original problem, but now it is much easier to work out the solution. See if you can find the minimum solution in 16 moves.

An old puzzle that also lends itself to analysis by the string-and-bead technique makes use of the star diagram shown in Figure 8. To work on this puzzle you need seven pennies or small counters.

The problem is this: Place a penny on any point of the star and move it along a black line to a different point. Once the penny is moved it must remain on the point to which you moved it.

Now place a second penny on any unoccupied point of the star, and move it in similar fashion to any other unoccupied point. Continue in this way until all seven pennies have been placed on points.

8

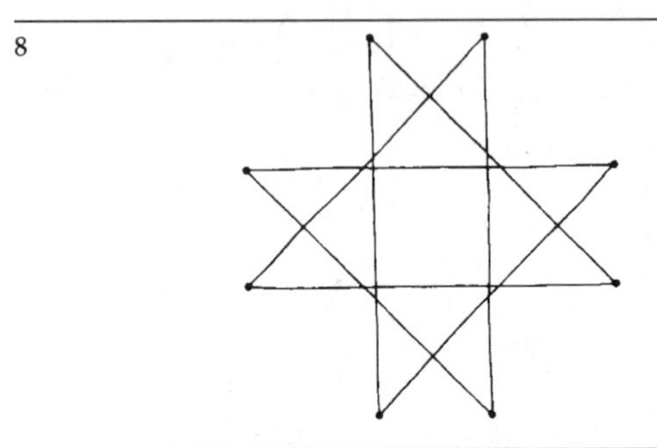

You will soon discover that unless you place and move the pennies according to a carefully designed plan, you will find yourself trapped in a position that will not permit you to continue. The problem is to devise a system for placing and moving all seven pennies according to the rules. Can you work out the system?

The star graph can be opened out like the graph of the first two knight problems into one that is circular. It is now easy to place and move all seven counters. There are many ways to do it. One simple system is to make any move you like with the first penny. Thereafter, always place and move the next penny so that it ends on the spot vacated by the previously placed coin.

Try this puzzle on your friends. Very few of them will be able to solve it even after you have demonstrated (rapidly) how it can be done.

Surprising Swords

奇妙的刀

Study this picture carefully. Do you see anything wrong with it?

Look at the sword. It cannot possibly fit into its scabbard.

If these two swords have a uniform cross section they will fit into a scabbard of the corresponding shape. But can you think of a third shape for a sword and its scabbard?

Did you have the insight to think of three dimensional curves? It turns out that a spiral curve called the helix is the only other possible shape for a sword and its scabbard.

General Helix

螺旋线在生物学、化学与物理学中具有十分重要的地位。它与直线和圆等平面曲线的重要差别在于它有“手性”，在物理学中称为自旋。由于手性，显示出许多生物模式，特别像贝壳花纹的不同，有许多不对称问题由此而来。

The helix has become an important structure in modern science, especially in biology and nuclear physics. It is the structure of the DNA molecule. Unlike its one-and two-dimensional cousins, the straight line and the circle, the helix has a "handedness"; that is, it can be right-handed or left-handed. A straight line or a circle is identical with its mirror image, but a helix is not. In the mirror it "goes the other way," as Lewis Carroll's Alice said when she looked at the room behind the looking-glass. A neutrino, for example, travels with the speed of light, but because it has "spin" it traces (in a sense) a helical path in space-time. Neutrinos and antineutrinos have helices of opposite handedness.

There are many examples of helices in nature and in everyday life. A right-handed helix is traditionally defined as one that coils clockwise as it "goes away" from you. Screws, bolts and nuts are usually right-handed. Helical structures such as circular staircases, candy canes, springs, and the helical strands of ropes, cables and strings, come in both forms. Do barber poles?

Examples of spirals in nature include the horns of many animals, conical sea shells, the long tooth of the narwhal, the cochlea of the human ear, and umbilical cords. In the plant world helices turn up in stalks, stems, tendrils, seeds, flowers, cones, leaves, tree trunks and so on. Squirrels trace helices when they run up and down a tree. Bats fly in helical paths when they emerge from a cave. Conical helices are exhibited by such weather phenomena as whirlpools and tornadoes. Water flows helically down drains. For more examples of helices in nature see *The Ambidextrous Universe* by Martin Gardner.

A regular helix is a curve that coils around a circular cylinder, making a constant angle with the cylinder's elements. (The elements are straight lines on the surface that parallel the axis.) Call this constant angle theta. It is easy to see that if theta is zero, the helix is a straight line. If theta is 90 degrees, the helix is a circle. This can be established analytically by using the parametric equations for a helix and letting theta vary between 0 and 90 degrees. Thus, the straight line and the circle are limiting forms of the more general space curve called the helix. The regular helix is the only space curve of constant curvature and torsion. This explains why the helix and its two limiting forms provide the only shapes for the swords and their scabbards.

One projection of the helix on the plane is obviously a circle. If a projection is made at right angles to the axis of the helix, it is a sine curve. Again, this is easily verified by examining the curve's parametric equations. Indeed,this furnishes a pleasant introduction to the sine curve and its properties.

Here is an amusing story problem involving a helix that has a good aha! solution. A cylindrical tower, 100 meters high, has an inside elevator. Around the tower's outside is a helical stairway that has a constant theta angle of 60 degrees with the vertical. The tower's diameter is 13 meters.

One day Mr. and Mrs. Pizza rode the elevator to the observation deck at the top of the tower. Their son, Tomato Pizza, climbed the stairway all the way from the bottom to the top. When he arrived at the observation deck he was breathing heavily.

"No wonder you're beat, son," said Mr. Pizza. "You must have gone four times the distance we did. And you did it all on foot."

"You're wrong, Dad," said Tom. "I only went *twice* as far."

Who was right. Tom or his father? One is inclined to think it necessary to use the diameter of the circular tower in calculating the length of the helical staircase. Surprisingly, the tower's diameter of 13 meters is extraneous information that can be ignored entirely!

The reason the diameter is irrelevant is that the helical stairway corresponds to the hypotenuse of a right triangle with angles of 60, 30 and 90 degrees, and a height of 100 meters. The hypotenuse of such a triangle is, of course, twice the height (the side opposite the 30-degree angle). Therefore Tom was right.

You can verify this by unwinding a cardboard mailing tube, or the tube around which paper towels come. The outcome may astonish you. You will see at once that the tube's helical edge has a length that is independent of the width of the cylinder into which the right triangle is rolled.

Payoff at the Poles

航空飞行

这是一个有趣的问题。一般人容易忽略的地方在于靠近南极之处可以 100 千米绕一圈、两圈、三圈乃至无穷!

Bet-a-dollar Dan, a famous gambler, is having a drink with his friend Dick, an airline pilot.

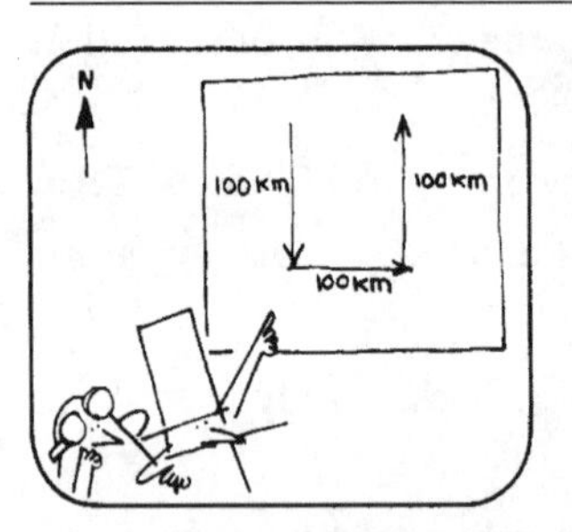

Dan: Dick. I'll bet you a dollar that you can't figure this one out. A pilot flies due south 100 kilometers, then goes east 100 kilometers, then north 100 kilometers and finds that he's right back where he started from. Where did he start from?

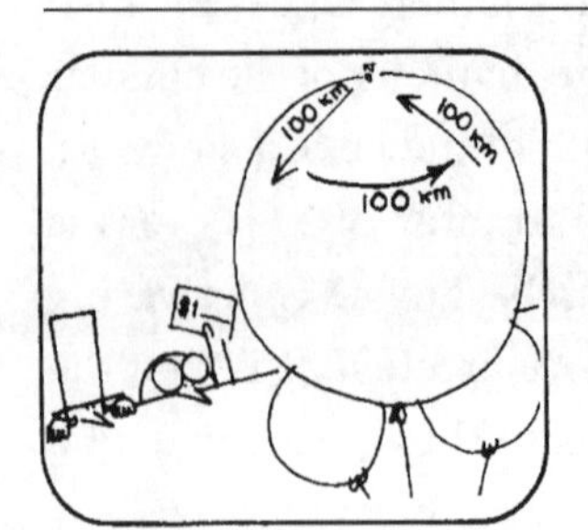

Dick: I'll take that bet. Dan. That's an old one. He started from the North Pole.

Dan: Right. Here's your dollar. Now, I'll bet you another dollar that you can't think of a different starting place.

Dick thought about it for a long time.

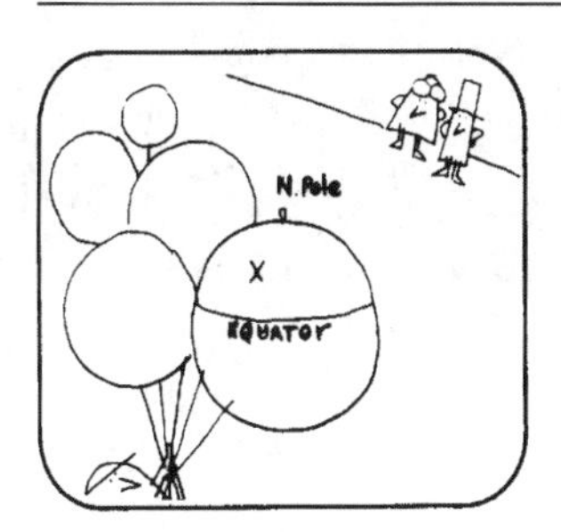

Dick: There can't be another starting place, Dan, and I can prove it. Suppose the pilot starts anywhere between the North Pole and the Equator.

Dick: It's obvious that he can't get back to where he started. And if he starts on the Equator, he'll end up about 100 kilometers from his starting spot.

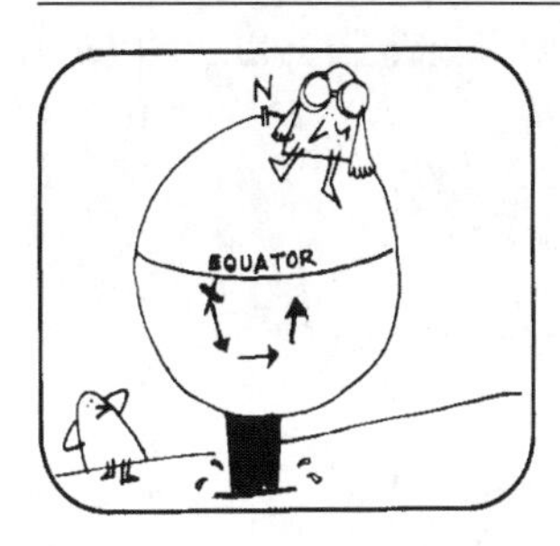

Dick: And by starting off anywhere south of the Equator he'll miss his starting spot by way more than 100 kilometers.

Dan: Okay, want to bet double or nothing that there's no other possible starting spot?

Dick took the bet and lost. Do you see why?

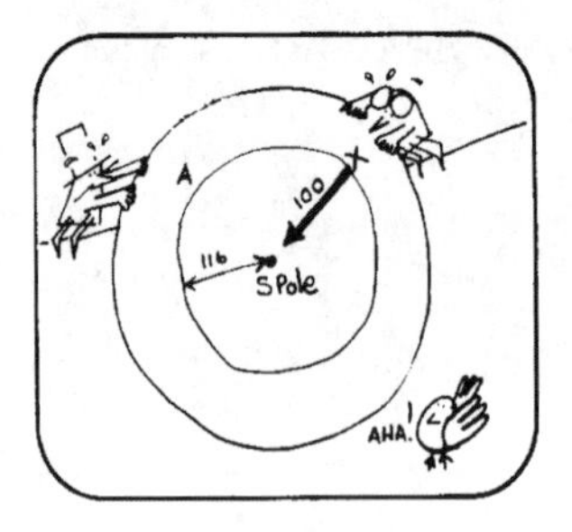

Suppose the pilot starts anywhere on circle A which is 116 kilometers from the South Pole. Then he flies south 100 kilometers.

Now when he goes 100 kilometers east he has made a complete revolution around the pole. So when he goes north 100 kilometers he has to be back where he started from. Right?

Dick: Right. Here's your $ 2.
Dan: Wanna bet another dollar that I can't find still another starting spot?
Dick: You mean that's not at the North Pole and not on circle A?
Dan: That's what I mean.

Dick: Alright then, make it $50.

Poor Dick lost again. What aha! did he miss this time?

Starting Spots

The insight that Dick missed when he lost the second bet is this. A pilot can start from a spot so near the south pole that when he flies 100 kilometers east he goes *twice* around the pole instead of just once as in the previous solution. This introduces a new circle, each point of which is a solution to the original problem. Similarly, the pilot can start anywhere on a still smaller circle so that his eastern flight takes him three times around the pole, or four times, and so on for every natural number. It turns out, therefore, that the starting points that solve the problem lie on an infinite set of concentric circles. The circles all have their centers at the south pole, and radii that approach 100 as the limit.

Here is a different navigational problem that involves a fascinating curve on the sphere known as a loxodrome or rhumb-line. A pilot starts at the equator and flies due northeast. Where will his flight end? How long is the path and what does it look like?

当航行不是朝正南或正北而是与径线成固定角，则得到斜驶线，它在球面上表示为环绕北极的无穷次的螺线。有趣的是它的总长度却是有限的。也就是它的长度可表为收敛的无穷级数。

You may be surprised to find that the path is a spiral that cuts the earth's meridians at a constant angle and ends precisely at the north pole. The path is a spherical helix that "strangles" the north pole, but only after making an infinite number of circles around it. Think of the pilot as a moving point. Paradoxically, even though the point goes around the pole an infinite number of times, the path has a finite length that can be calculated. Thus, if the pilot (represented by a point) travels at a constant speed, he or she reaches the north pole in a finite length of time.

A loxodrome, plotted on a flat map, has different forms depending on the type of map projection. On the familiar world map called the Mercator projection, it is plotted as a straight line. Indeed, this is why a Mercator map is so useful to navigators. If a ship or plane travels in a constant compass direction, the path is a straight line that is easy to draw on the map.

What happens if a pilot starts at the North pole and

flies due southwest? This is a reversal of the previous problem. The path is a loxodrome as before, but now we cannot specify the spot where it will reach the equator. It can meet the equator at *any* spot. You can prove this by time reversal. Just start the plane at any spot on the equator and its backward flight must carry it to the north pole. However, if the pilot continues on his forward path beyond the equator, his loxodrome will strangle the south pole.

When a loxodrome is projected on a plane parallel to the equator and tangent to a pole, it is an equiangular or logarithmic spiral. This is a spiral that always cuts its radius vector at a constant angle.

The four-bug problem is another well-known path problem, also involving a logarithmic spiral, but with a beautiful aha! solution that avoids a lot of laborious calculation. We give it here with a story line about the Pizza family and its pet turtles.

Tom Pizza has trained his four turtles so that Abner always crawls toward Bertha, Bertha toward Charles, Charles toward Delilah, and Delilah toward Abner. One day he put the four turtles in ABCD order at the four corners of a square room. He and his parents watched to see what would happen.

"Very interesting, son," said Mr. Pizza. "Each turtle is crawling directly toward the turtle on its right. They all go the same speed, so at every instant they are at the corners of a square" (see Figure 9).

9

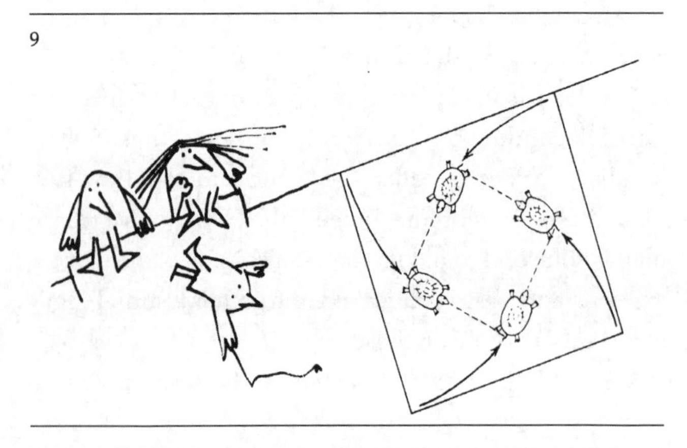

"Yes, Dad," said Tom, "and the square keeps turning as it gets smaller and smaller. Look! They're meeting right at the center!"

Assume that each turtle crawls at a constant rate of 1 centimeter per second, and that the square room is 3 meters on the side. How long will it take the turtles to meet at the center? Of course, we must idealize the problem by thinking of the turtles as points.

Mr. Pizza tried to solve the problem by calculus, using his new pocket programmable calculator. Suddenly Mrs. Pizza shouted: "You don't need calculus. Pepperone! It's simple. The time is 5 minutes."

What was Mrs. Pizza's insight?

Consider two adjacent turtles, say Abner and Bertha. At every instant Bertha is moving at right angles to Abner, who is pursuing her, because Abner always crawls directly toward her while she is crawling directly toward Charles. This is why the turtles are at all times at the corners of a square. Since Bertha's movement is never toward Abner or away from him, her motion neither adds nor subtracts from the distance between herself and Abner. Her motion, therefore, becomes irrelevant. It is the same as if Bertha remained in her corner and Abner crawled toward her along the side of the square room.

The above insight is the key to the solution. Abner's curved path must have exactly the same length as the side of the square. Since the side is 300 centimeters, and Abner crawls at 1 centimeter per second, it will take him 300 seconds, or 5 minutes, to reach Bertha. The same is true of the other three turtles. At the end of 5 minutes. all four turtles meet at the square's center.

四臭虫或四龟问题说明可以不必计算来推出正确的结果。关键在于每一个龟的爬行曲线长度正好等于正方形的边长。

With the help of a pocket calculator, it is not difficult to diagram the paths of the turtles in small increments of time, drawing the four sides of the square at the end of each interval. The result is a startling pattern (see Figure 10).

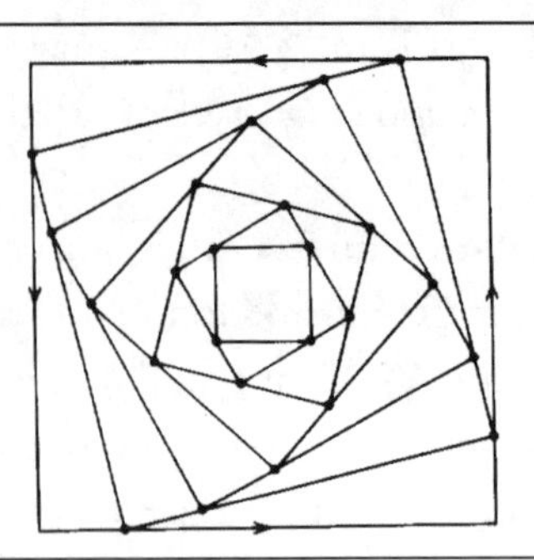

Can you generalize the problem to the corners of all regular polygons? Investigate first the equilateral triangle, then the pentagon. Can you find a general formula for the lengths of the pursuit paths, given the length of the side of the starting polygon? What happens in the limiting case when an infinite number of turtles (points), starting at the corners of an infinite-sided polygon, chase each other? Will they ever meet? Suppose the initial polygons are not regular. What happens if four turtles start at the corners of a rectangular room that is not square?

Suppose that after the four turtles meet at the square's center, in our original problem, they find that they dislike one another so they crawl outwards, each moving directly away from the turtle on its left. Will the turtles necessarily return to the four corners of the room?

Quibble's Matches

奎贝尔的火柴

火柴游戏是最重要的一类数学游戏，极富启发性。

Mabel is showing a match puzzle to Professor Quibble.

Mabel: Move just two matches and end up with four of the same sized squares. And don't break, overlap, or double up the matches.

Prof. Quibble: That's an old timer, Mabel. All you have to do is move these two matches.

Professor Quibble then took away 4 matches, leaving 12 on the table.
Prof. Quibble: Alright now Mabel, make six unit squares with these 12 matches.
Mabel had to give up. But maybe you can help her out.

Match Games

The insight Mabel needed to solve Professor Quibble's match puzzle is that she has not been told that the matches must remain on the plane. By going into a third dimension, the 12 matches form the 12 edges of a unit cube, which, of course, has 6 square faces. It is an insight similar to Rosie's when she found a way to cut the cheese.

这个游戏再次提示到第3维去的思想。最明显例子是用6根火柴拼成4个正三角形.下面6题各有巧招。例如，第1个问题，一般认为至少移动2根火柴才能得到一个大正方形。但仔细观察一下，只要把右边的火柴稍稍向右移一下就得出一个小正方形。

A better known version of the same problem is to form four identical equilateral triangles with six matches. The solution in this case is to form the skeleton of a regular tetrahedron.

Here are six other clever match or toothpick puzzles that have aha! solutions. Can you do them?

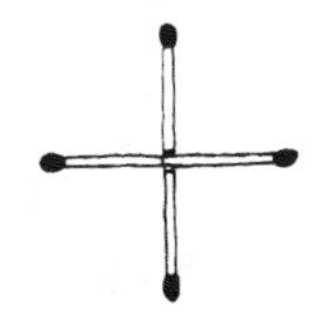

1. Move the smallest number of matches to make a square.

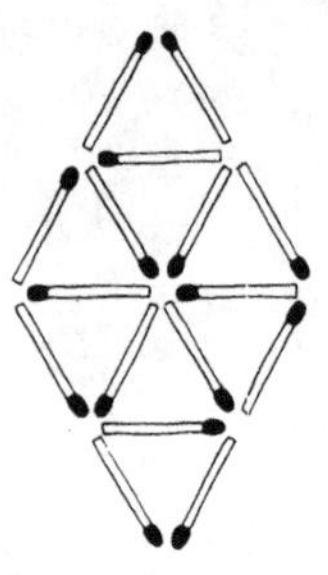

2. Remove the smallest number of matches to leave four equilateral triangles of the same size as the eight shown. There must be no loose ends.

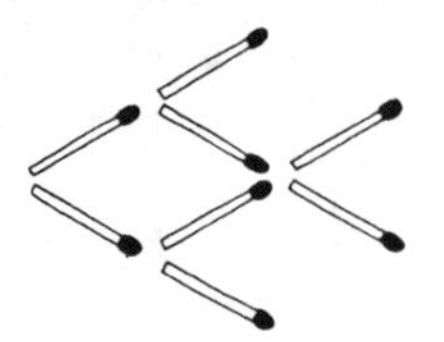

3. Move the smallest number of matches to make the fish swim the opposite way.

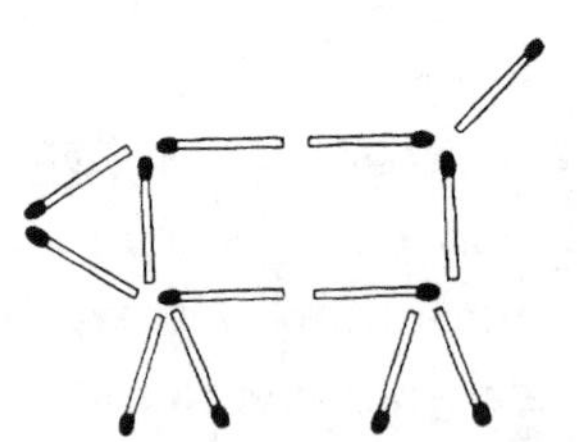

4. Move the smallest number of matches to make the pig look the opposite way.

5. Move the smallest number of matches to get the cherry outside the old-fashioned glass. The glass may have any orientation at the finish, but, of course, the cherry cannot be moved.

6. Move the smallest number of matches to get the olive outside the martini glass. As before, the glass at the finish may be turned any way you like, but the olive must not be moved.

It would spoil most of the fun if we printed the actual solutions, so instead we will give only the correct minimum numbers:

1. One.
2. Four.
3. Three.
4. Two.
5. Two.
6. None.

Devilish Divisions

巧妙的划分

Ransom is a surveyor who specializes in dividing curious shaped lots into congruent parts.

这是剖分理论第二组问题，即把一个图形剖分成几块相等的图形。这三个问题是最简单的。

On one occasion he was asked to cut this lot into four identical regions. How do you think he did it?

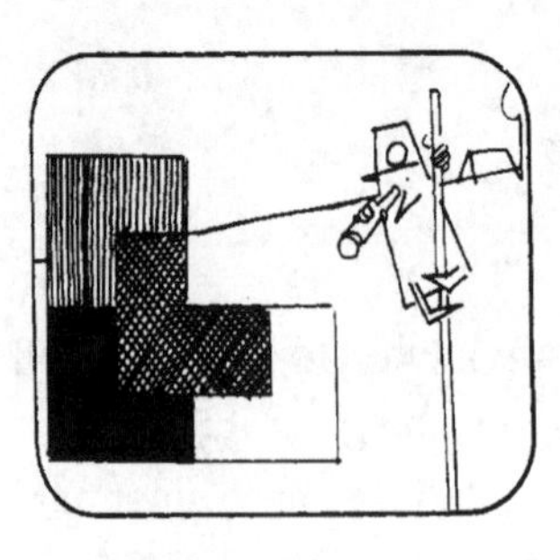

This is the only way it can be done.

Ransom's next job was to cut this piece of land into four congruent parts. And it wasn't easy to do so.

However, he persevered and finally found one solution.

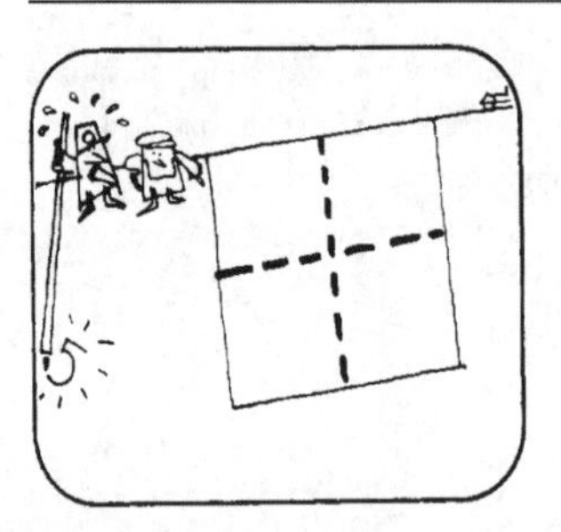

Dividing square lots into four identical regions offered no challenge to Ransom, but when he was asked to divide a square lot into five congruent regions he was puzzled.

Ransom: I don't understand it.
There has to be away. Hmm.
Aha! I see it now.
Can you figure out what Ransom's insight was?

Ransom: It's ridiculous. The same method could be used to divide a square into any number of congruent pieces.

Dissection Theory

Ransom's three problems form a series that is great fun to try on friends. The first two puzzles are solved with strange shapes. These shapes subtly suggest that the square, since it cannot be cut into five squares, must be cut into five curious shapes. It is surprising how few people can think of the obvious solution. Incidentally, it is the *only* way a square can be cut into five congruent shapes.

当图形比较复杂时，问题就需要动点脑筋了。这些问题在数学上很重要，涉及铺砌问题和结晶学问题。物理学家 Roger Penrose 用两种样板得出五重对称性，1984 年发现准晶具有五重对称性，开创了全新的领域。

After you have caught a friend with this puzzle, you can probably catch him or her a second time with a closely related fourth problem. First show your friend how the field in Figure 11 can be divided into four congruent shapes. Can this field also be cut into *three* identical parts?

Your friend will probably soon give this up as much too difficult. He or she will be dumbfounded when you show how it can be solved easily with exactly the same insight that enabled Ransom to cut a square into five identical regions. The answer is given in Figure 12. As before, the technique obviously permits one to cut the field into any number of identical regions.

Puzzles of this type, as well as the puzzles related to our cheese cutting problem, belong to a colorful branch of recreational mathematics sometimes called dissection theory. They provide valuable insights into the solution of many practical problems in plane and solid geometry. Ransom's first two problems are especially interesting because each field is cut into pieces of the same shape as the original field. If this can be done, the shape is called a *rep-tile*.

Figure 13 shows several other rep-tiles. Can you cut each of these into congruent shapes that replicate the original shape? It is clear that if you have an infinite supply of any rep-tile, you can tile the entire plane in a nonperiodic way. For example, consider the L-shaped rep-tile that is the first field solved by Ransom. Four such pieces make a large L-tile, then four larger L-tiles make a still larger L-tile, and this process can be continued to infinity to tile the infinite plane. Note also that we can go to infinity in the opposite direction by cutting each tile into four smaller L-tiles, and those in turn can be cut into still smaller L-tiles, and so on *ad infinitum*.

Not much is known about rep-tiles. All known rep-tiles also tile the plane periodically. That is, they tile the plane in such a way that there is a fundamental region of the pattern that tiles the plane by translation, without rotation or reflection. Is there a reptile that will not also tile periodically? This is an outstanding unsolved question of tiling theory.

Even less is known about solid rep-tiles. The cube is, of course, such a figure because eight cubes go together to make a larger cube, just as four squares go together to make a larger square. Can you think of any other examples of a solid rep-tile?

If the congruent shapes are not required to be similar to the field that is dissected, many other unusual puzzles can be devised. Figure 14, for example, is a T-shape formed by five unit squares. It cannot be cut into four smaller T's, but can you cut it into four congruent regions of some other shape?

Even the task of dissecting a plane figure into as few as *two* congruent parts can be difficult. Figure 15 shows some examples that you may enjoy solving. The solutions are shown in the back of the text.

Another elegant branch of dissection theory has to do with cutting a given polygon into the smallest number of pieces, of any shape, that can be rearranged to make a different polygon that also is specified. For example, into how few pieces can a square be cut that will fit together to make an equilateral triangle. (The answer is four.) This field is beautifully covered in *Recreational Problems in Geometric Dissections & How to Solve Them* by Harry Lindgren.

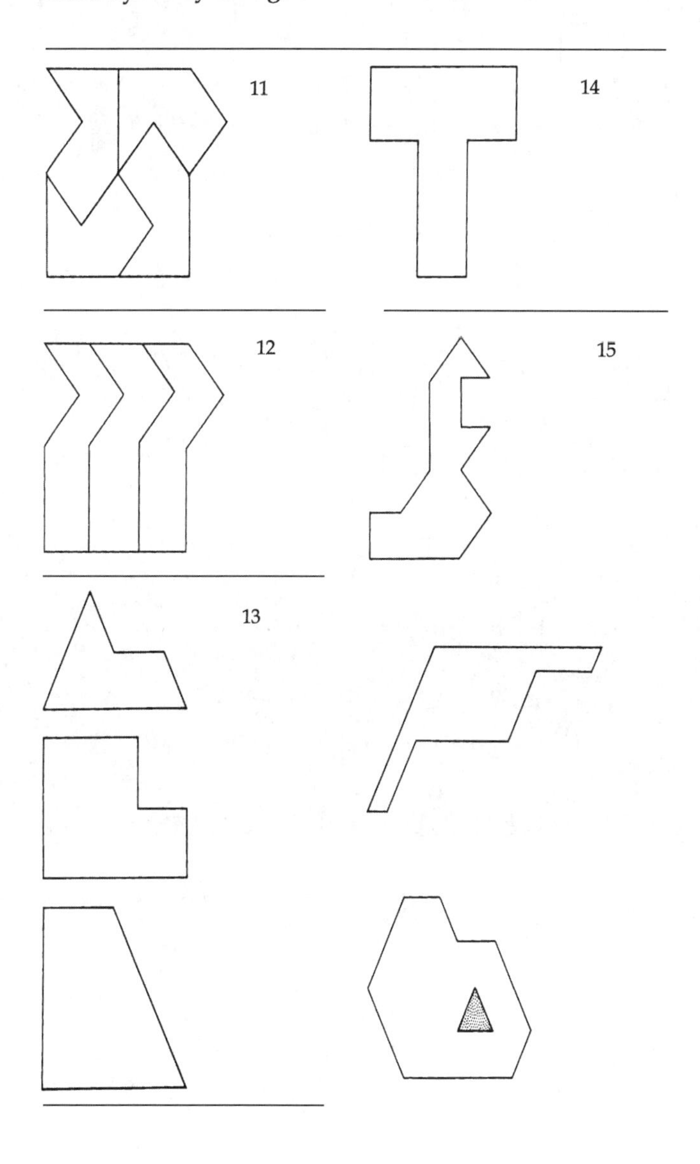

15 (continued)

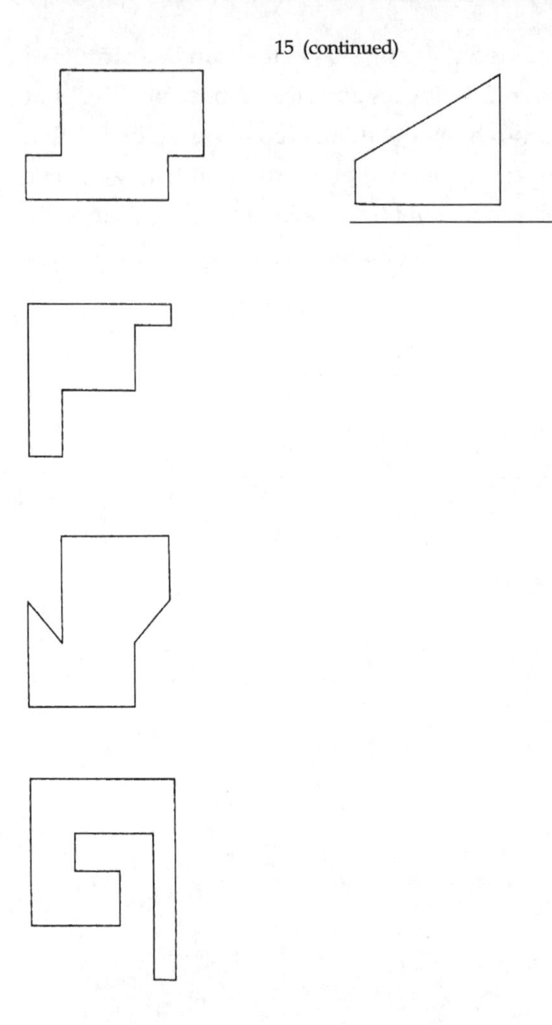

Miss Euclid's Cubes

殴几里得小姐的立方体

这是剖分理论另一系列。

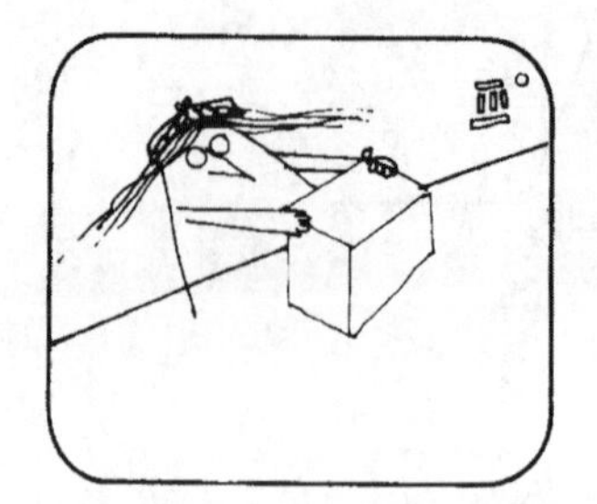

Miss Euclid put a large wooden cube on her desk.

Miss Euclid: I have a very practical test for you today. Just three questions about this cube.

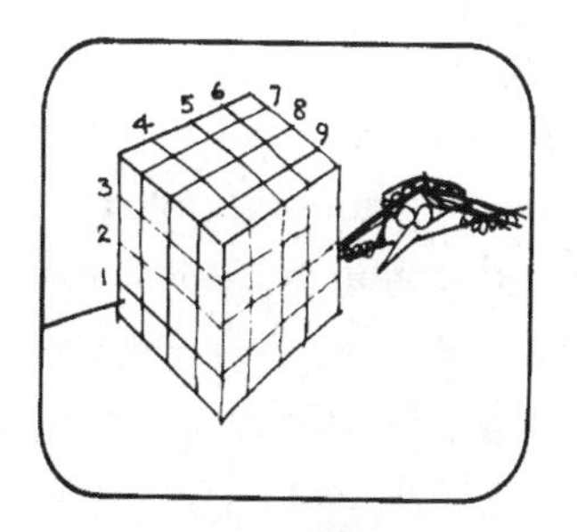

Miss Euclid: If we have a table saw we can cut this cube into 64 unit cubes by making nine cuts.

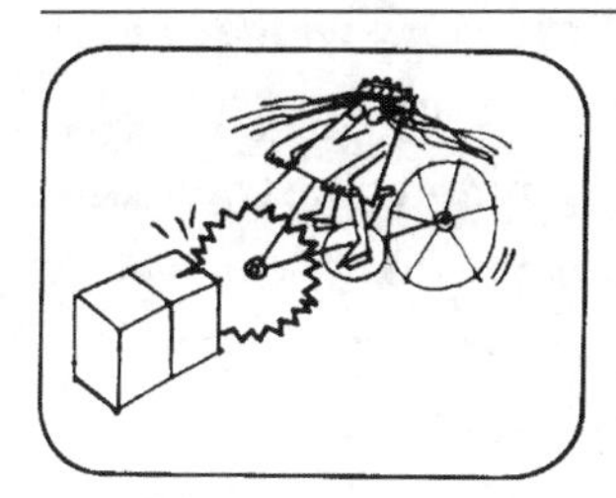

Miss Euclid: And if we're allowed to rearrange the pieces before each cut, we can do it with only six cuts. Your first question is to prove that you can't do it in fewer than six cuts.

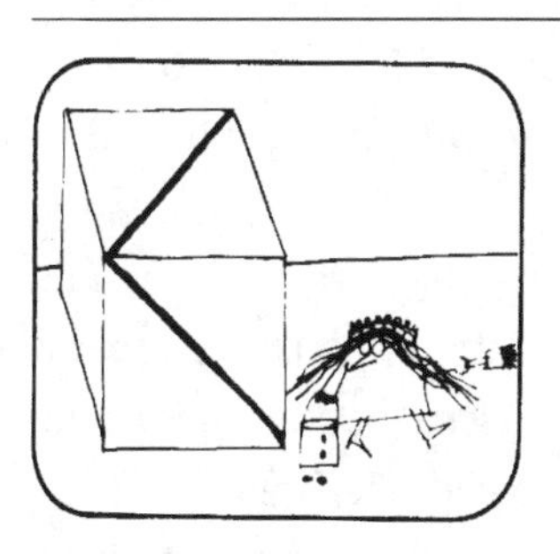

While the students were working on the first problem Miss Euclid drew a diagonal on two faces of the cube so that there was a common vertex.

Miss Euclid: Your next problem is to find the size of the planar angle formed by these two diagonals and their common vertex.

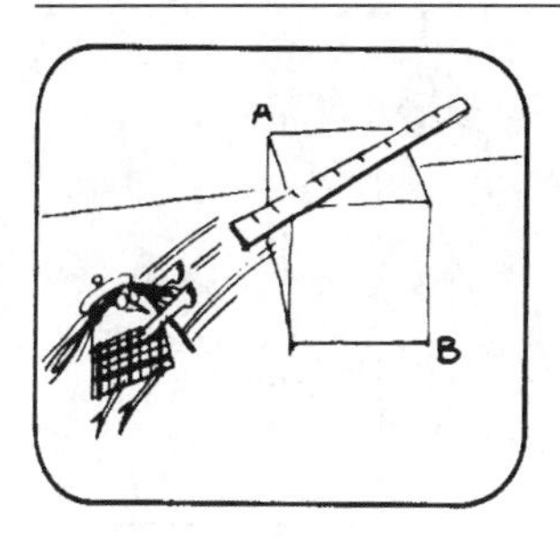

Miss Euclid prepared for the last question by placing a meter stick on top of the cube.

Miss Euclid: What's the simplest way to use this ruler to measure the length of the space diagonal from A to B?

How did you do on the test? I got two out of three.

Miss Euclid's Solids

Solution to Problem 1: To prove that a 4×4×4 cube cannot be sliced into 64 unit cubes with fewer than six planar cuts (allowing for rearrangement of pieces after each cut), just consider any of the 8 cubes in the interior. Since none of these cubes has a face on the outside of the large cube, each of its six faces must be cut with one planar cut. Since no plane can cut more than one of the cube's faces at a time.

这些问题培养立体观念。例如，由正六边形把立方体一分为二。这些是立体几何很好的补充。计算一个立体上两点间距离的最好办法是把它们放在同一平面上。

clearly at least six cuts are necessary for the six faces.

Is there a systematic general procedure for slicing any rectangular parallelepiped with integral sides into unit cubes with a minimum number of planar cuts, allowing rearrangements of pieces between cuts? Yes, the method is as follows. Along each of three edges that meet at a corner, determine the minimum number of cuts necessary to slice the cube through that edge to make unit-wide sections. This minimum number is obtained by dividing the edge as nearly in half as possible, then putting the two pieces together and repeating this procedure until the unit-wide sections are obtained. The sum of these three minimums, one for each edge, is the answer sought.

For example, a $3 \times 4 \times 5$ block requires 7 cuts: 2 for the 3 side, 2 for the 4 side, and 3 for the 5 side, or 7 in all. A proof of this algorithm was first published in *Mathematics Magazine* in 1952.

Solution to Problem 2: The insight that solves this problem is seeing that a third diagonal can be drawn on another face of the cube that will join the free ends of the two diagonals drawn by Miss Euclid; see Figure 16.

16

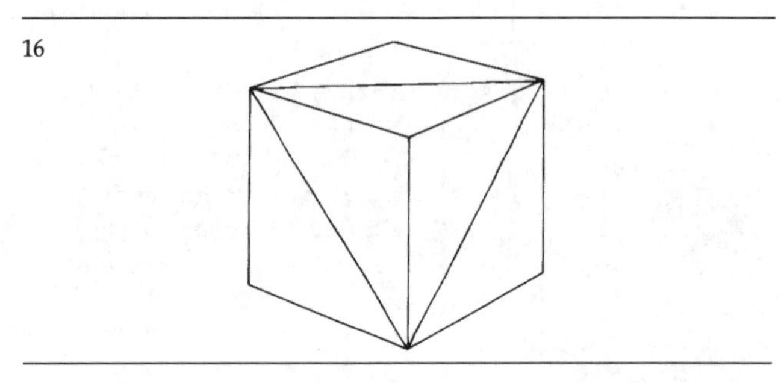

These three lines form an equilateral triangle. Since each angle of such a triangle is 60 degrees, we have proved that the angle on Miss Euclid's cube is 60 degrees.

There is an elegant extension of this problem. Suppose Miss Euclid draws two lines on a cube as shown in Figure 17, joining three midpoints of three edges. What is the size of the obtuse planar angle made by the two lines?

The solution is obtained as before. First, continue

17

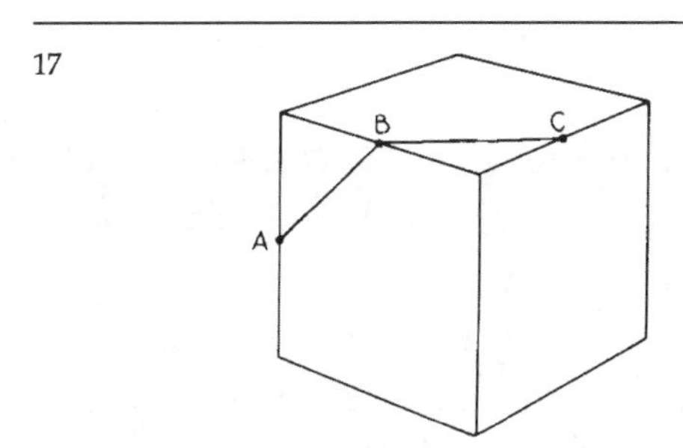

the lines by joining midpoints on the other four faces so that the six lines make a closed path around the cube. It is evident that the six line segments are of equal lengths, and also that any adjacent pair makes the same angle. The six lines, therefore, outline a regular hexagon if we can show that the vertices are all coplanar. This may take a bit of deduction or analytic geometry, but you can convince yourself it is true by actually sawing a wooden cube into two identical parts along a plane that cuts the cube through the six midpoints.

The fact that a cube can be cut in half so that the cross section is a regular hexagon is quite surprising and almost counterintuitive. Of course, once we know that the two original lines are a pair of adjacent sides of a regular hexagon, we know that they make an angle of 120 degrees.

Figure 17 suggests another interesting problem. Suppose a fly wants to crawl on the cube's surface from midpoint A to midpoint C. Is the path traced by the two line segments the shortest route the fly can take?

Here the insight is to recognize that the shortest path from A to C can be found by "unfolding" the cube so that two adjacent sides are flat, then drawing a straight line on the surface from A to C. Now we must be careful because there are two ways to do this: Unfold the front and top faces, or unfold the front and rightmost faces. The first case gives a path of length $\sqrt{2}$; the second case gives a path of the length $\sqrt{2.5}$. This proves that the path shown in Figure 17 is indeed the shortest path on the cube's surface from A to C.

Solution to Problem 3: Of course. you can measure a

第 3 个问题对学数学的人有很多启发。数学教材只教你如何通过公式来算，而在实际问题中往往没有现成公式，而需要进行实地测量。例如，求一张不规则形纸片的面积，可以用称重法解决。

side of the cube, then apply the Pythagorean theorem twice to obtain the space diagonal. But a much simpler method is to place the cube flush with the corner of a rectangular table. Place a small mark on the table's edge that is distance x from the table's corner, where x is the cube's side. Now slide the cube along the edge of the table to the other side of the mark, as shown in Figure 18. The distance from *A* to *B* is obviously the same as the cube's space diagonal, and it can be measured directly with the ruler.

18

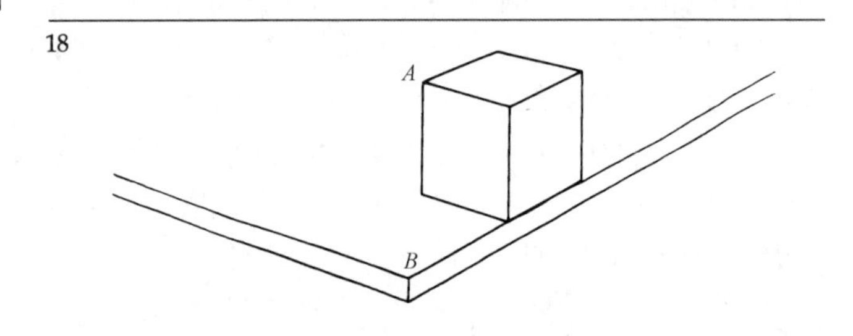

How would you measure the radius of a large sphere if you had only a ruler that was about 2/3 the sphere's diameter? One simple method is to smear a bit of soot or lipstick on a portion of the sphere, then place the sphere on the floor and push it against the wall so that the soot or lipstick marks the wall at the spot where the sphere touches it. The height of this spot, easily measured with the ruler, is the sphere's radius. Can you think of similar ingenious ways to measure the heights of cones and pyramids? How can you measure accurately the radius of a cylindrical pipe with a carpenter's square?

Carpet Confusion
地毯难题

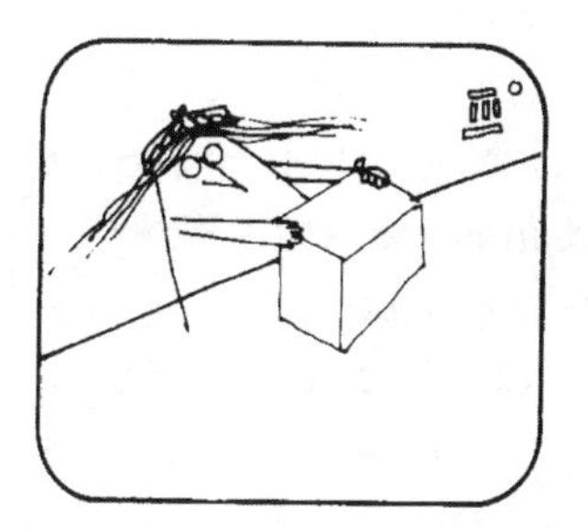

The Tack Carpet Company was asked to provide wall-to-wall carpeting for a ring shaped corridor in a new airport.

这是平面几何一个很好的综合问题。

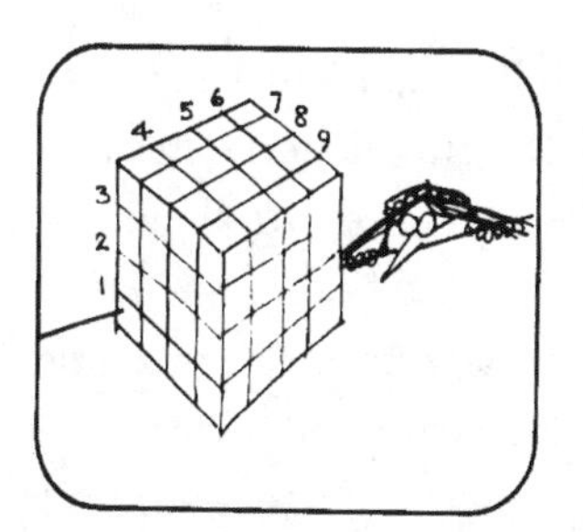

When Mr. Tack saw the plans he was angry. The only measurement given was the length of a chord that was tangent to the inner wall.

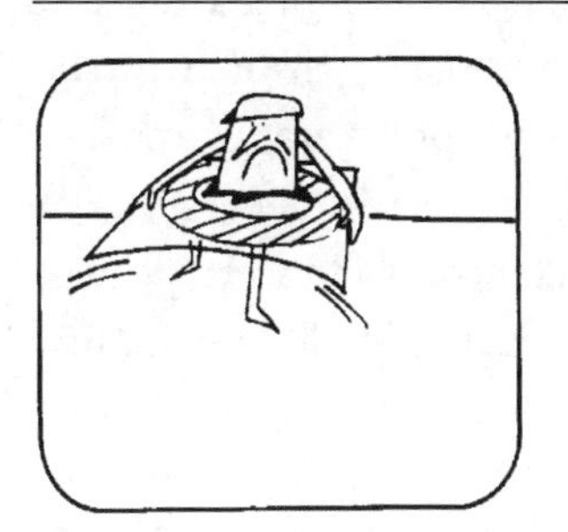

Mr. Tack: Confound it. How can I give them an estimate of carpet cost when I don't know the area of that blue ring between the two circles? I'd better go and see my designer, Mr. Sharp.

Mr. Sharp, a skilled geometer, wasn't too upset.

Mr. Sharp: That chord is the only length I asked for, Mr. Tack. I just plug it into a formula I have and it gives the ring's area.

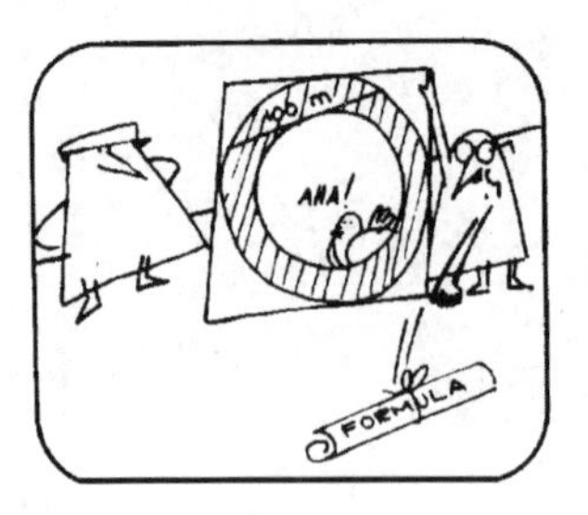

Mr. Tack looked surprised for a minute, then smiled.

Mr. Tack: Thank you, Mr. Sharp, but I don't need you or your formula. I don't have to know the areas of the two circles either. I can give you the result immediately.

Do you know how Mr. Tack did it?

这是一个惊人的定理。它也教你处理问题时先考虑最简单的极限情形。

Astonishing Theorem

Mr. Tack reasoned as follows. I know that Mr. Sharp is a skilled geometer, therefore there must indeed be a formula for the area of the ring when one is given only the length of the chord tangent to the inner circle. Put another way, the radii of the two circles can be any two numbers as long as the length of the chord remains 100 meters.

Mr. Tack then asked himself what happens when the radius of the inner circle is reduced to zero. Its minimum length. In this case the ring degenerates to a circle, the diameter of which is the chord of 100 meters. The area of this circle is pi times 50^2 or close to 7,854 square meters. Assuming the existence of a formula, this must also be the area of the ring between the circles.

In general, the area of any ring is equal to the area of a circle whose diameter is the longest straight line that can be drawn inside the ring. This astonishing theorem can be proved easily by using the formula for the area of a circle.

A 3-dimensional analog of this problem is that of determining the volume of a section of cylindrical pipe with thick walls, when given only the length of the longest line that can be drawn on one of the pipe's ends (see Figure 19). This 1ine corresponds to our tangent line, from which we can

19

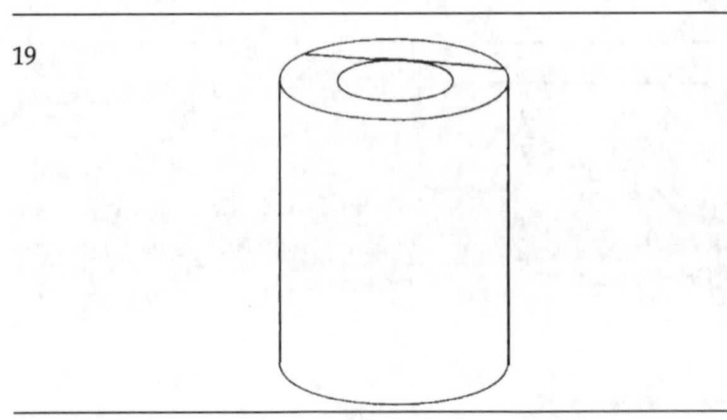

quickly determine the area of the ring on the pipe's end. Multiplying this by the pipe's length gives the pipe's volume.

A less obvious analog is the following beautiful problem. A cylindrical hole 6 centimeters long is drilled straight through the center of a solid sphere. What is the volume that remains? Again, it seems impossible to determine the volume without more data. However, it can be shown, without calculus, that the volume of the sphere that remains is always the same as the volume of a solid sphere whose diameter is equal to the hole's length.

As before, this result is immediately obtained on the assumption that the problem can be solved! If there is a solution, the volume of the sphere that remains after the hole is drilled must be independent of the hole's diameter. So—we reduce the hole's diameter to zero, its lowest limit. The hole degenerates to a straight line that is the diameter of a solid sphere. The answer, therefore, is $(4/3)\pi 3^3 = 36\pi$ cubic centimeters.

The Curious Cake Cut

蛋糕的稀奇切法

这是第四类的剖分问题：等面积分割问题。

Mr. Jones is finishing dinner with his wife, a teenage son, and a seven-year-old daughter, Susan.

It was Susan's birthday and Mrs. Jones had baked a small square cake. It was 20 centimeters by 20 centimeters and 5 centimeters high. Thick icing covered the top and four sides.

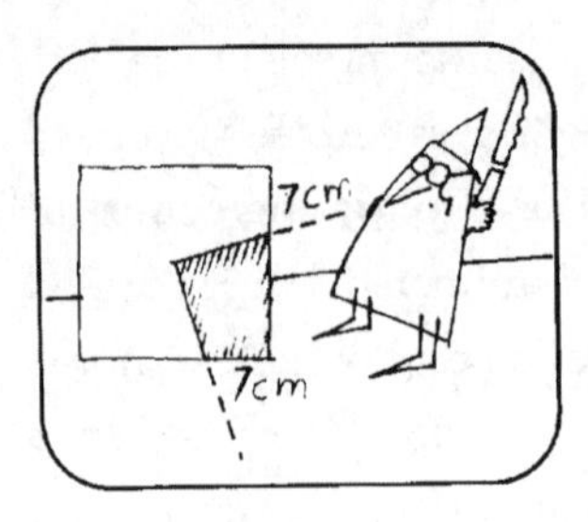

Mr. Jones: What a lovely cake, my dear. Just enough for all of us. I'll cut Susan's piece first, and since she's just turned seven. I'll start each cut seven centimeters from a corner and cut to the center.

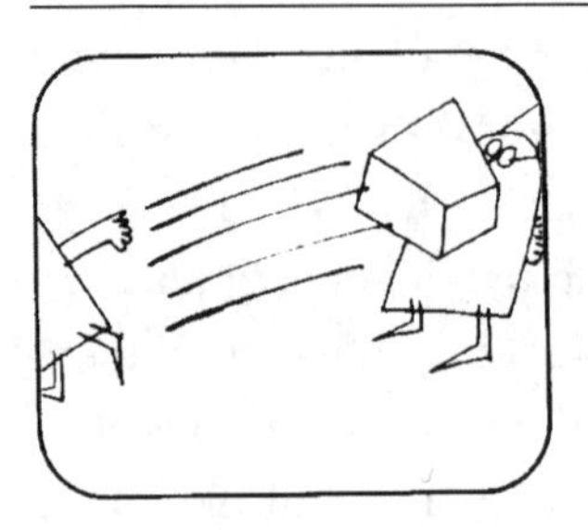

It was a strange shaped piece. And it wasn't long before Susan started to complain.

Susan: You didn't give me enough, Daddy. That's not one quarter of the cake. And even if it were I didn't get enough icing.

Her brother disagreed.

Susan's Brother: You're too greedy, Susan. I think that Dad gave you too much and that you should give some back.

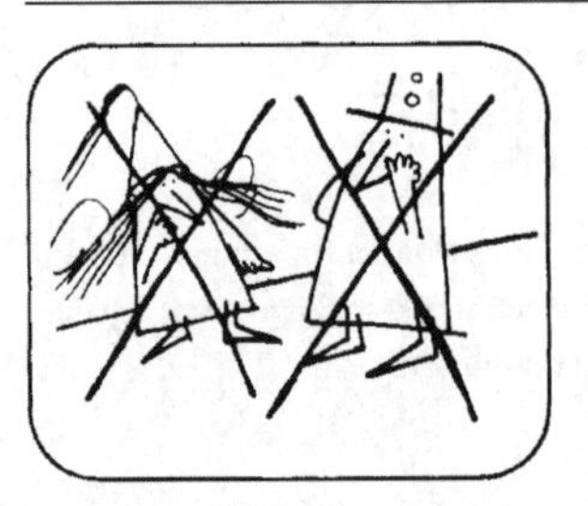

Mr. Jones: Well you're both wrong. The piece is exactly one-fourth the volume of the cake and it also has exactly one fourth of the icing on it. Can you explain why Mr. Jones said this?

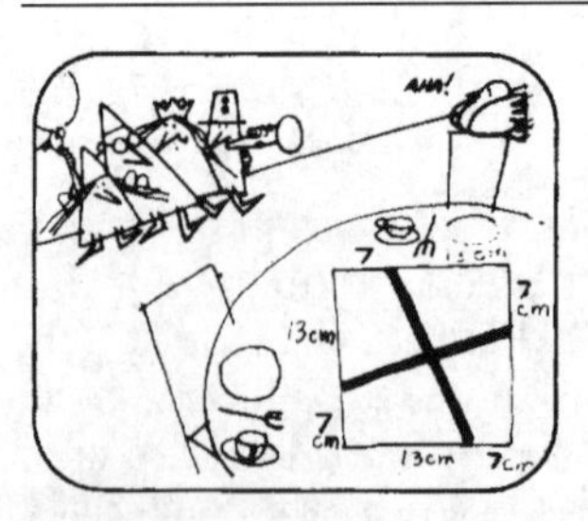

All that you have to do is to extend the two cuts past the center of the cake to the other side. Now it's clear why the lines cut the cake into four congruent parts. Isn't it?

Cake Cutting

The cake cutting problem generalizes easily to all other

regular polygons. For example, suppose a cake has the shape of an equilateral triangle, and that two cuts are made from the center at an angle of 360/3 = 120 degrees as shown in Figure 20. The piece is clearly one-third of

最有趣的是图 21 和图 22 的行数和重组问题。

20

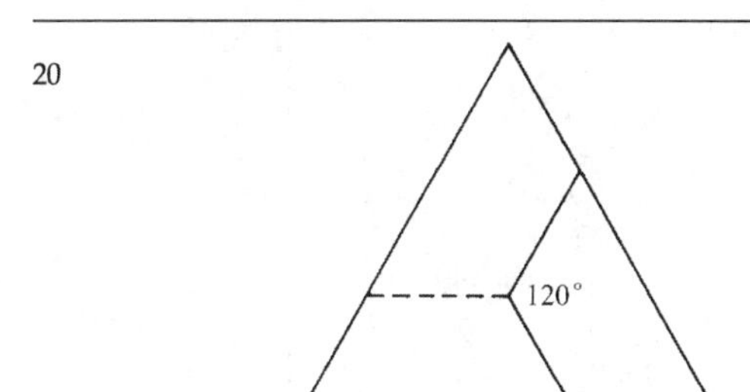

the cake, as we can see by drawing the dotted line. If the cake is pentagonal, two cuts at 360/5 = 72 degrees give one-fifth of the cake. If the cake is hexagonal, two cuts at 360/6 = 60 degrees give one-sixth of the cake. This generalizes to all higher polygons, although the angle is not always integral as in the cases given above.

The dissection of the square into four congruent pieces, as shown in Figure 21, has for decades been a popular dissection puzzle. If you give friends the four pieces of such a dissection, cut from a square of cardboard,

21

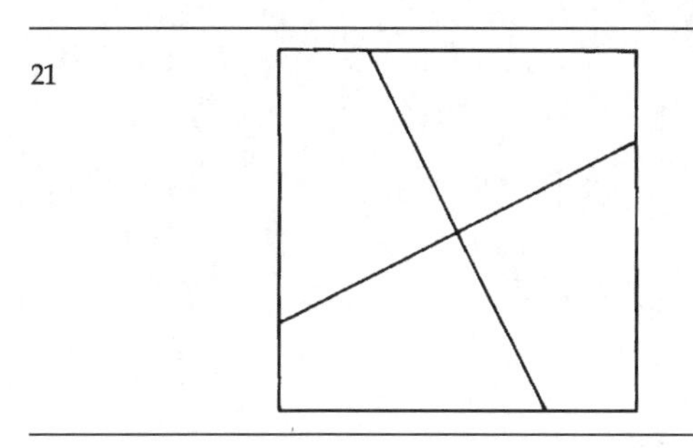

and ask them to make a square, they usually find it difficult. After they have solved the puzzle, ask them to use the same four pieces to form *two* squares.

This is something of a swindle because it can be done only if one has the aha! that the second square is a hole at the center of another square as shown in Figure 22. The size of this hole depends on the angle that each cut makes with the side of the original square. If the angle is zero, the hole is zero. If the angle is 45 degrees, the hole reaches its maximum size.

22

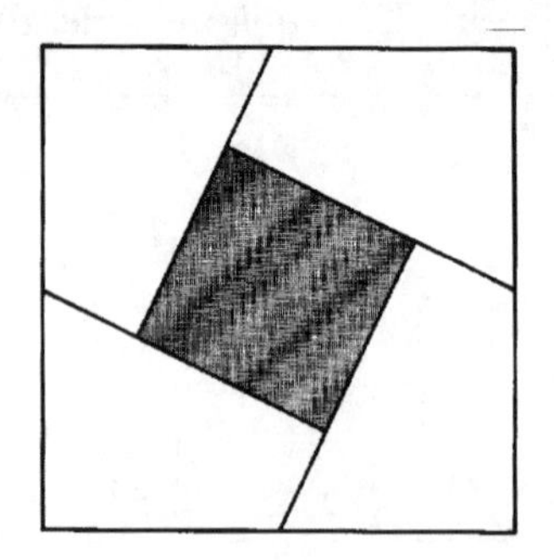

Chapter 3

Number aha!

数　字

Puzzles about arithmetic

Arithmetic can be defined in many ways. Here we confine arithmetic to be the study of the integers and the results of operating on the numbers by addition, subtraction, multiplication, and division.

arithmetic一般有两种译法：一种是“算术”，也就是计算的技术和方法；另一种是“数论”，也就是关于数的性质的一般理论，其结果往往以定理的形式来陈述。有些陈述，如哥德巴赫猜想，虽然意思明白易懂，但至今尚未得到完全的证明。数论是数学中的一个前沿领域，还有许多猜想与问题尚未解决。

At some time in the childhood of the human race (no anthropologist knows when), primitive humans somehow slowly discovered that things can be counted, and that it does not make a difference in what order the counting is done. If you count two sheep on your fingers, it does not matter which sheep you start with, or whether you start the count on your thumb or your little finger. You always end on 2. And if you count two sheep, then another, you always get 3.

Awareness of such arithmetical theorems as 2 + 1 = 3 must have slowly evolved over many centuries. If we could see a motion picture of the past, we probably would not be able to find a single century about which we could say: “This is when the human race discovered arithmetic.” Children slowly become aware of numbers in the same vague way. There may be a time when a child first says “One plus one is two”, but the child may be aware of the meaning of this sentence long before he or she verbalizes it.

All the true theorems of arithmetic follow at once from the axioms and definitions of the number system, but that does not mean that we can recognize the truth or falsity of an arithmetical statement just by hearing it. If someone declares that 12, 345, 679 times 9 equals 111, 111, 111, you may not believe it until you prove it by doing the multiplication. And there are theorems in arithmetic that are simple to state but so deep that nobody knows if they are true or not. Goldbach's conjecture is a famous example. Is every even number (greater than 2) the sum of two primes? No one has yet proved that the answer is “yes” or found a counterexample.

In this section, we consider a variety of simple problems about counting numbers, all of which have easy solutions if they are properly approached. We have tried to select problems that, although very elementary, introduce important concepts and techniques that lead into deeper levels of what used to be called the "higher arithmetic" and is now called "number theory". "Broken Records", for example, introduces Diophantine analysis: The finding of integral solutions to equations. "One too Many" involves the all-important concept of the lowest common multiple, and leads to a magic trick based on the valuable "Chinese remainder theorem".

Binary sorting, so important in computer search and sort theory, underlies the technique for guessing Helen's unlisted phone number, and introduces the binary system of notation. The "pigeon hole principle", fundamental to many deep proofs in number theory, is invoked in proving two amusing results: One about dollar bills, the other about hairs on the head. The fact that two integers are "relatively prime" (have no common divisors) provides a surprisingly quick way of proving that the hour, minute, and second hands of a watch are never together except at 12 o'clock—a theorem usually proved by tedious algebra.

A problem about counting bottles uses modulo arithmetic to obtain an easy solution. This leads to the "Josephus problem", a classic number problem that can be modeled in an exciting way with a deck of playing cards.

Although the puzzles in this section are what mathematicians consider trivial, they open up paths of exploration into branches of number theory that are far from trivial. And they cannot fail to impress you with the elegance and richness of that oldest of all deductive systems, the system that manipulates the symbols for the familiar counting numbers.

Broken Records
掰开的唱片

Bob and Helen are enthusiastic puzzle buffs. Their favorite pastime is trying to stump each other and their friends with puzzling questions.

正整数是数学中最早研究的对象。许多事物必须以整数来计算，如唱片和金鱼。不过在考虑有关它们的问题时，借用一下分数是方便的，只要前提和结果是合理的整数即可。

As Bob and Helen went past a record store, Bob said.

Bob: Do you still have your country western records?

Helen: No, I gave half of them, and half a record more, to Suzy.

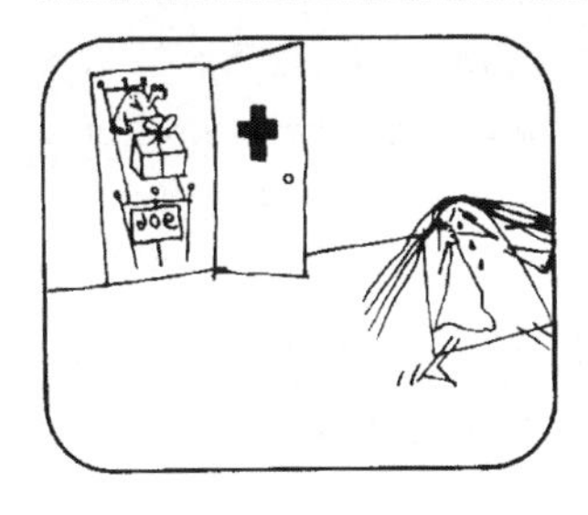

Helen: Then I gave half of what was left, and half a record more, to Joe.

Helen: That left me with just one record. And I'll give it to you if you can tell me how many country western records I had to begin with.

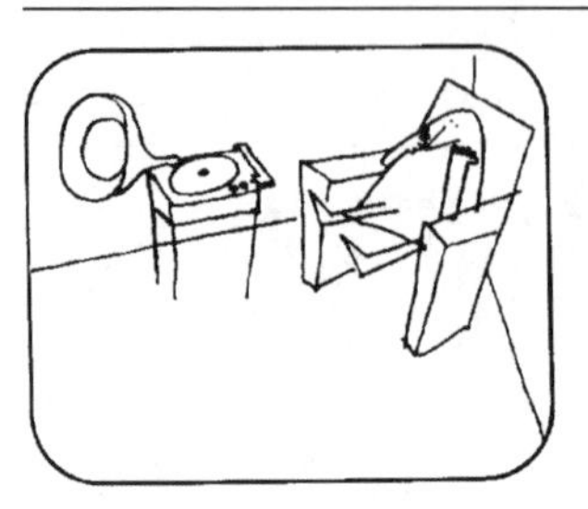

Bob was puzzled because he couldn't see how half a record could be of any use.

Suddenly he had an aha! and realized that not a single record had been broken. He answered Helen's question and she gave him the last record. What insight did Bob have?

Half Wholes

这些问题都可用列出方程再求解的方法解决，但如果通过灵机一动的方法就得考虑到出现 1/2 的情形总是来源于奇数，而且出现奇数次 1/2 再加上 1/2，又可以还原成整数，而不必真的打破唱片。

Did you fall into the trap of thinking that half of something plus ½ can't be a whole number? If so, you probably tried to solve the problem by thinking of broken records and quickly became lost. The aha! insight is the realization that half an odd number of records, plus half a record, is a whole number.

Since only one record remained after Helen's last gift, she must have had three records before she gave one to Joe. Half of 3 is 1½ , and 1½ + ½ = 2, so Helen's last gift was 2 records. This left her with one whole record at the finish. It is now easy to work backward and see that she must have started with seven records and given 4 to Suzy.

The problem can, of course, be solved algebraically, and writing and solving the equation for it is an excellent exercise in elementary algebra. It is surprising that such a simple little problem has the complicated equation:

$$x-\left(\frac{x}{2}+\frac{1}{2}\right)-\left[\frac{x-\left(\frac{x}{2}+\frac{1}{2}\right)}{2}+\frac{1}{2}\right]=1$$

By varying the parameters, it is easy to make up new problems of the same type. For example, assume that Helen follows the same procedure of giving away at each step half her records plus half a record, but does this three times instead of twice, and ends with no records at all. How many did she have at the start? You may be amused to discover that the answer is the same as before: 7 records! The third step consists in giving away the entire last record. How many records does she start with if she follows the halving procedure four times and ends with a single record? Five times? What kind of sequence is generated by these numbers?

The fraction given away each time may also be varied. Suppose Helen at each step gives away a third of her records plus one third of a record, and after two steps finds she has three records left. How many did she start with? Is there a solution if she follows this thirding procedure three times and ends with three records? You will find that by varying the parameters—number of steps, fractional amount, and number of whole records at the finish—there are not always solutions in the sense that no record need ever be broken. Under what restraints can problems of this type be devised that never require breaking a record?

There also is no need to have the fractional amount the same at each step. Here, for instance, is a puzzle in which the fraction varies:

A boy has the hobby of breeding goldfish. He decides to sell all his fish. He does this in five steps:

1. He sells one half of his fish plus half a fish.

2. He sells a third of what remains, plus one third of a fish.

3. He sells a fourth of what remains, plus one fourth of a fish.

4. He sells a fifth of what remains, plus one fifth of a fish.

He now has 11 goldfish left. Of course, no fish is divided or injured in any way. How many did he start with? The answer is 59 fish, but the problem is not as easy to solve as the previous ones. See if you can work it out.

Here is a somewhat different problem of the same general kind.

A lady has a certain number of dollar bills in her purse. She has no other money.

1. She spends half the money on a hat, and gives a dollar to a beggar outside the store.

2. She spends half the remaining dollars for lunch, and tips the waiter two dollars.

3. She spends half the remaining dollars for a book, then before she goes home she visits a cocktail lounge where she spends three dollars on drinks.

She now has one dollar bill left. Assuming that she never changed a dollar bill, how many bills did she start with?

The answer appears at the back of the book.

Note that in all these variations we are told the number of items that are left at the finish. Without this information the problem often can still be solved, but it may require the solving of in determinate equations in integers. The most famous problem of this type was the basis of a short story by the American writer Ben Ames Williams that appeared in the *Saturday Evening Post*, October 9, 1926.

The story, titled *Coconuts*, tells about five men and a monkey who were shipwrecked on an island. They spent the first day gathering coconuts. During the night, one man woke up and decided to take his share of coconuts. He divided them into five piles. One coconut was left over so he gave it to the monkey, then hid his share and went back to sleep.

Soon a second man woke up and did the same

thing. After dividing the coconuts into five piles, one coconut was left over which he gave to the monkey. He then hid his share and went back to bed. The third, fourth and fifth men followed exactly the same procedure. The next morning, after they all woke up, they divided the remaining coconuts into five equal shares. This time no coconuts were left over.

How many coconuts did they originally gather?

The problem has an infinite number of answers, the lowest of which is 3, 121. It is not an easy problem.

Speaking of coconuts taken from a pile, here is a "quickie" that may stump you momentarily: If 25 coconuts are piled up in a jungle clearing, and a monkey steals all but 7, how many coconuts will be left? The answer is not 18.

Loch Ness Monster

海峡怪兽

Bob: If the length of the Loch Ness Monster is 20 meters and half its own length, how long is it?

这问题十分平凡，关键是确切地理解题意。

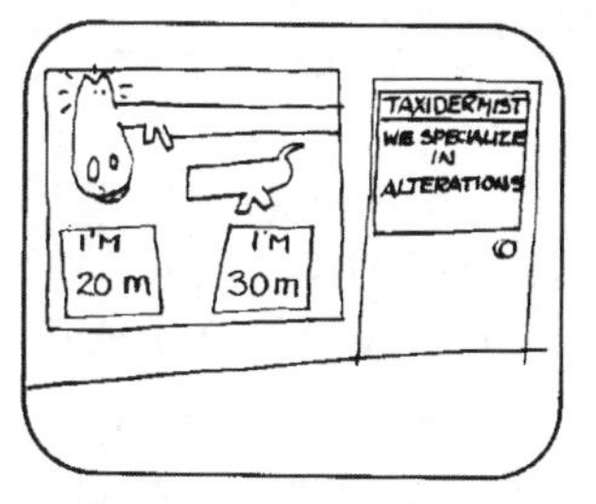

Helen: Let's see. Twenty and half of twenty is thirty. So it's 30 meters long.

Bob: Helen, I'm surprised at you. You've contradicted yourself. How can it have a length of 20 meters, and also a length of 30 meters?

Helen: You're right. The only way the sentence makes sense is if the total length is the sum of 20 meters and half the length. It's simple enough now. Can you figure out how long the monster is?

Half a Length?

Bob's phrasing of the problem is as follows: The monster's length is equal to the sum of 20 meters and half the monster's length. Imagine the monster is divided into two equal lengths. If the monster's length is the sum of one of these halves, plus 20 meters, then 20 meters must be the *other* half. Therefore the total length is 40 meters.

The algebraic equation is simple. If x is the total length, then:

$$x=20+x/2$$

Now that you see how ridiculously simple the solution is, how quickly can you solve the following variant? A brick on one pan of a balance scale exactly balances with three-quarters of a brick and three-quarters of a kilogram on the other side. How much does the brick weigh?

The Loch Ness Monster problem illustrates the importance of understanding *exactly* what a question means before trying to answer it. If your first interpretation of a problem leads to a contradiction, then either the question has no answer or you have not correctly understood the problem.

One Too Many

多余的一个

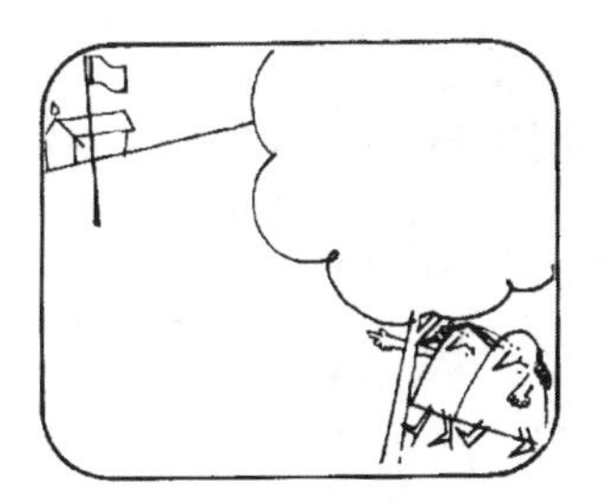

While Bob and Helen were crossing the park they saw the Nixon High School Band practicing for a parade.

由余数求全数是数学游戏中一类有趣问题，也是初等数论最基本的定理的来源。
这个问题首先是在中国古代约公元3世纪成书的《孙子算经》中所记载的："今有物不知其数，三三数之剩二，五五数之剩三，七七数之剩二，问物几何？答曰二十三。"这问题有多个名称，如"秦王暗点兵""物不知总""韩信点兵"等。

The band came marching by, four in a row, with one boy, poor Spiro, bringing up the rear. The band director was annoyed.

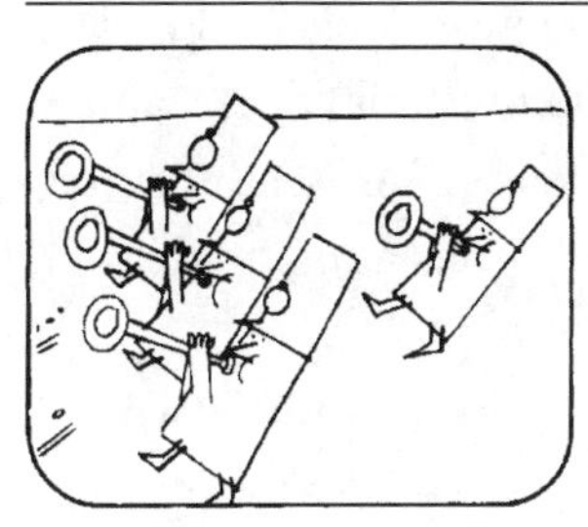

To eliminate that lonely musician in the back, the director told the band to march by in threes. But Spiro was still alone in the last row.

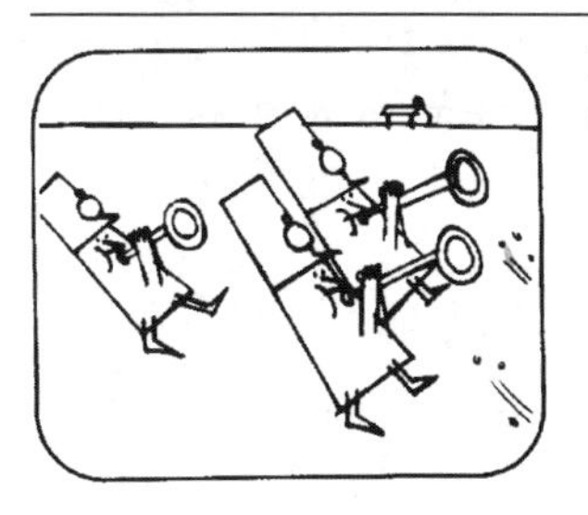

Even when the band marched by in twos, the same thing happened.

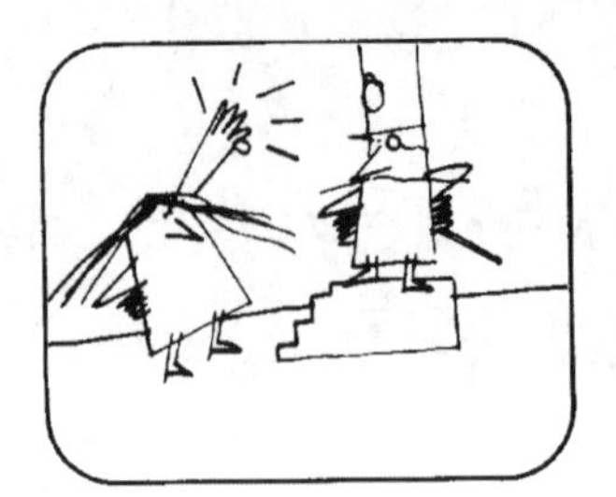

Although it was none of her business, Helen approached the director.
Helen: May I make a suggestion?
Bob: No, please go away, Fraülein, and don't bother me.

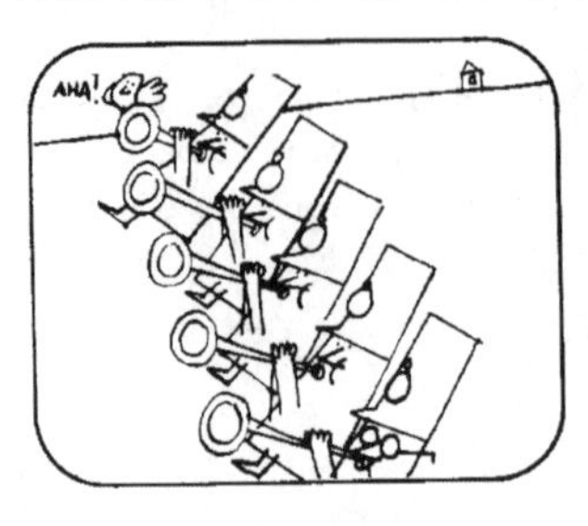

Helen: Well, I'll tell you anyway. Have them march by in fives.
Bob: My dear, I was just about to try five.

When the band marched by in fives all the rows were filled and Spiro wasn't alone any more. How many members did the band have?

Wholes From Remainders

对这类问题，中国古算的求法最后总结成为一个定理，称为“孙子定理”或“中国剩余定理”。本书前几个问题都具有同一余数，显然远为容易，只要求最小公倍数即可。但这也显示特殊的数学游戏问题经过推广可成为重要的数学定理。

Helen simply counted the number of players in the band and found that it was a multiple of five. But how can you, not seeing the entire band, determine the number of members?

Your aha! is this: The number has a remainder of 1, symbolized by Spiro, when divided by 2, 3 and 4. The smallest number with this property obviously is 1 greater than the LCM (lowest common multiple) of 2, 3 and 4. The LCM of these three divisors is 12. Any number that is one more than a multiple of 12 has a remainder of 1 when divided by 2, 3 and 4.

When the band marched by in fives, there was no remainder. Therefore, the number of persons must also be evenly divisible by 5. The numbers that solve the problem are the multiples of that appear in the following sequence:

13, 25, 37, 49, 61, 73, 85, 97, 109, 121, 133, 145, …

Since 145 is too large for a highschool band, the Nixon High School Band has either 85 or 25 members. We lack sufficient information to decide between these two answers.

A good variant of this problem is the same as before except that each time the band marched by in rows of 2,

3 and 4, the last row is one man *short*. How large is the band? Now we must write a sequence of numbers that are one *less* than multiples of 12 that are evenly divisible by 5. The sequence is: 35, 95, 155, …

The American puzzle maker Sam Loyd created the following more difficult variation. On St. Patrick's Day in New York City a large number of Irishmen were getting ready to march in the annual parade. The Grand Marshal tried arranging them in rows of 10, 9, 8, 7, 6, 5, 4, 3 and 2, but in every case there was a missing man in the last row. The men thought the gap was occupied by the ghost of Casey, who had died a few months before. Finally, in exasperation, the Grand Marshal ordered the men to march in single file. Assuming the number of men did not exceed 5,000, how many were there? This is a good exercise for finding the LCM of a set of numbers. The LCM in this case is 2,520. If we subtract Casey from this group, we have our answer: 2,519.

The problem seems to become more difficult if we are given a different remainder after each division, but this is not always the case. For example, consider this classic puzzle that goes back to Hindu arithmetic books of the seventh century.

A lady is carrying a basket of eggs. Frightened by a horse that gallops past her, she drops the basket and all the eggs break. When asked how many eggs the basket had contained, she replies by saying that she is very poor in arithmetic, but she remembers that when she counted the eggs by twos, threes, fours and fives, she had remainders of 1, 2, 3 and 4 eggs, respectively. How many eggs were originally in the basket?

This excellent problem seems at first to be more difficult than the previous ones. Actually, it is exactly the same as the first part of our second problem because in each case the remainder is one less than the divisor. So it is solved as before by finding the LCM and subtracting 1.

When the remainders have no uniform relation to the divisors, the problem does indeed become more complicated. Here is a clever pocket calculator trick based on a problem of this kind. Your friends will find it mystifying and intriguing.

The magician sits in a chair with his back to the audience. Someone thinks of any number not greater than 1, 000. He is asked to divide the number by 7 and call out the remainder, then divide the original number by 11 and call out the remainder, and finally to divide the original number by 13 and call out the remainder.

To speed the trick, someone in the audience determines the three remainders by using a pocket calculator. This is easy to do with the aid of the following algorithm: Perform the division, subtract the whole number part of the quotient, then multiply the result by the original divisor. Round the product to the nearest integer and you have the desired remainder.

The magician, knowing no more than the three remainders, is able to guess the chosen number. He does this by using his own pocket calculator and the following secret formula that he has on a small slip of paper pasted to the face of his calculator:

$$\frac{715a+364b+924c}{1{,}001}$$

In the formula, a, b and c are the three remainders in the order in which they are called out. The chosen number is the *remainder* after making the calculation given by the formula.

The strange looking formula is obtained as follows: The first coefficient is the lowest multiple of bc that is one more than a multiple of a. There are rules for finding this, but when the divisors are small, as in this case, it is easy to get the number by inspection. Simply go up the multiples of bc (143, 286, 429, 572, 715,…) until you reach a multiple that has a remainder of 1 when divided by a. In this case, $a = 7$, and the coefficient is 715.

The other two coefficients are obtained in the same way. The second one is the lowest multiple of ac that is one more than a multiple of b, and the third coefficient is the lowest multiple of *ab* that is one more than a multiple of *c*. The number below the line in the formula is simply $a \times b \times c$. In this way you can work ort a secret formula for any set of divisors provided that they are prime to one another (have no common divisors). It is not necessary that the divisors be primes themselves, as they are in our example.

The proof of the general formula involves modulo arithmetic and an understanding of a famous theorem called the Chinese Remainder Theorem. It is one of the most valuable of all number theorems, playing a basic role in many deep proofs as well as in the solution of scientific problems.

As an exercise, try working out the secret formula for a simpler version of the same trick-one that goes all the way back to Sun-tsu, a first-century Chinese mathematician for whom the Chinese remainder theorem is named. The chosen number is limited to numbers 1 through 105, and the divisors are 3, 5 and 7. The secret formula in this case is simple enough so that, with some practice, you can even do the calculations in your head.

Eyes and Legs

眼睛和脚

Before leaving the park, Bob and Helen walked through the zoo. In one enclosure they saw a mixture of giraffes and ostriches.

这个问题也很简单，在我国常称为“鸡兔同笼”问题。

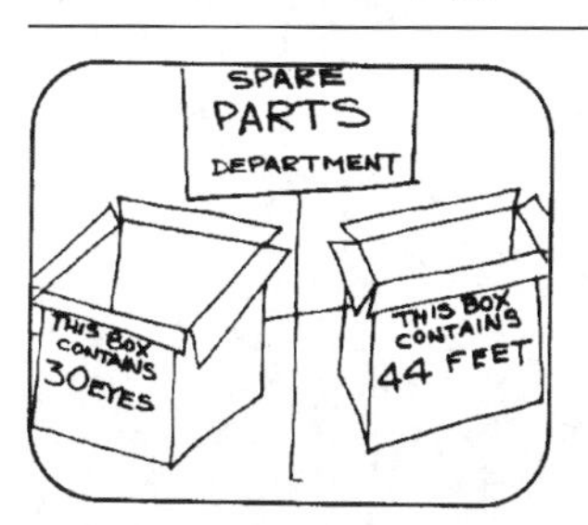

After they had left the zoo, Bob spoke to Helen.

Bob: Did you count the giraffes and ostriches?

Helen: No, how many were there?

Bob: You figure it out. Altogether they had 30 eyes and 44 feet.

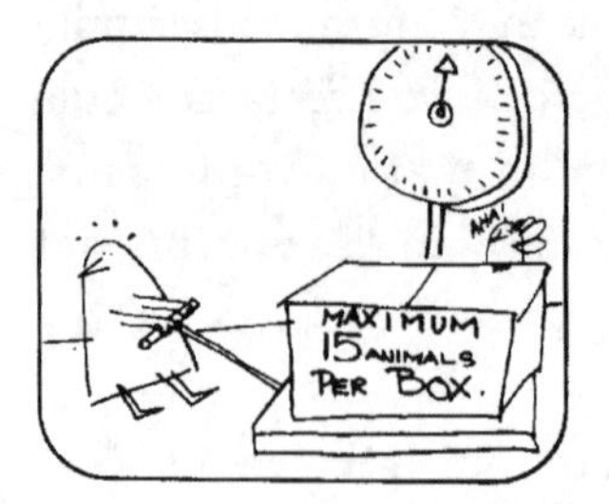

Helen's first aha! was to realize that thirty eyes meant 15 animals.

Helen: Now I can try all the possibilities, from no ostriches and 15 giraffes to 15 ostriches and no giraffes. But I don't need to do that.

Helen: If all 15 of the animals were to stand up on two feet there would be 30 feet on the ground .

Helen: But you said that there were 44 feet altogether. So 14 giraffe feet must be in the air. That makes 7 giraffes, right?

Bob: Right. And if there are 7 giraffes there have to be 8 ostriches.

Bipeds and Quadrupeds

The insight that solved this problem for Helen is easy to understand; however, you may wish to check the answer by algebra. Does your answer *agree*?

Here is an amusing follow-up puzzle that calls for a different kind of insight. A small circus has a certain number of horses and riders. Between them there are 50 feet and 18 heads. In addition, the circus has some jungle animals that have, altogether, 11 heads and 20 feet. There are twice as many four-footed jungle animals as there are two-footed creatures. How many horses, riders, and jungle animals are in the circus?

You should have little difficulty determining that there are 7 horses and 11 riders. But when you try to solve for the number of jungle animals, you may be surprised to find that you encounter a negative number.

Can you solve the problem before looking for the answer at the back of the book?

The Big Bump

撞车事件

When they reached Bob's sports car he offered to drive Helen home to her parent's new house.

这类“速度问题”或“追赶问题”都很简单。标准方法是列出代数方程。算术方法则是本书的“回溯法”。

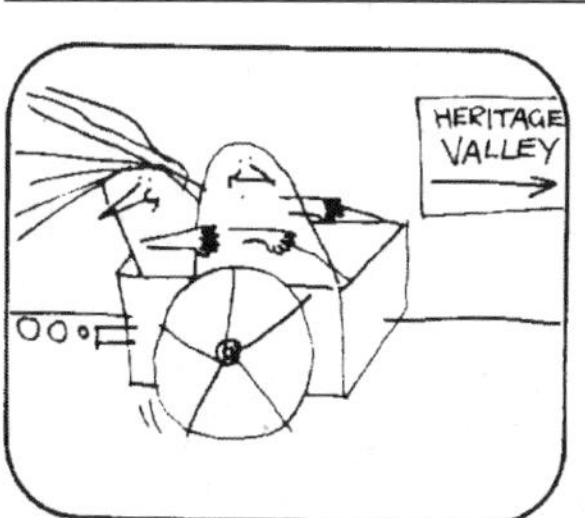

On the thruway, Bob thought of a good problem for Helen.

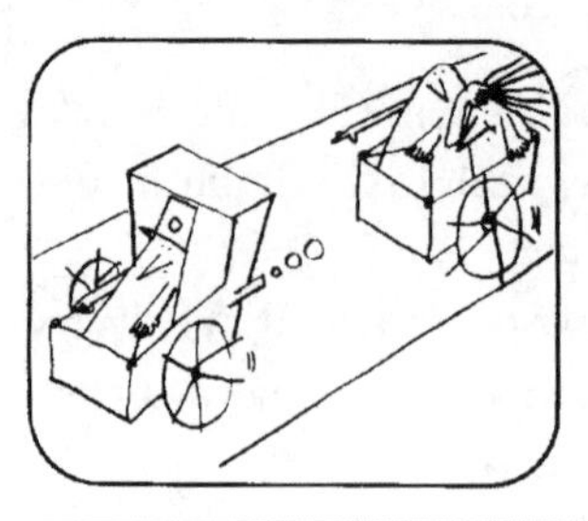

Bob: See that big truck ahead.
He's going pretty fast, but I'm gaining on him.

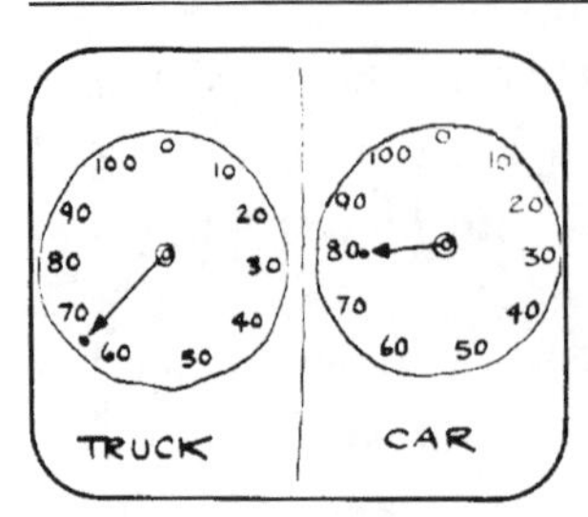

Bob: Now let's suppose that he's going a steady 65 kilometers per hour, and that I'm doing a steady 80.

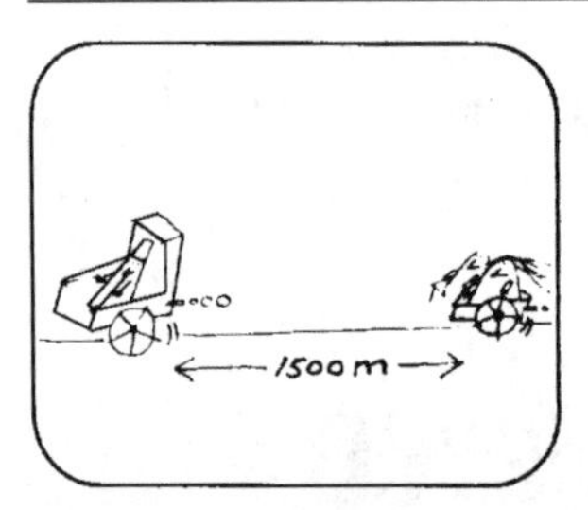

Bob: And let's say we're, 1,500 meters behind him right now.

Bob: So if we keep our steady speeds, and I don't pass him, we're sure to bump him. Your problem Helen, is to tell me how far apart we will be 1 minute before the crash.

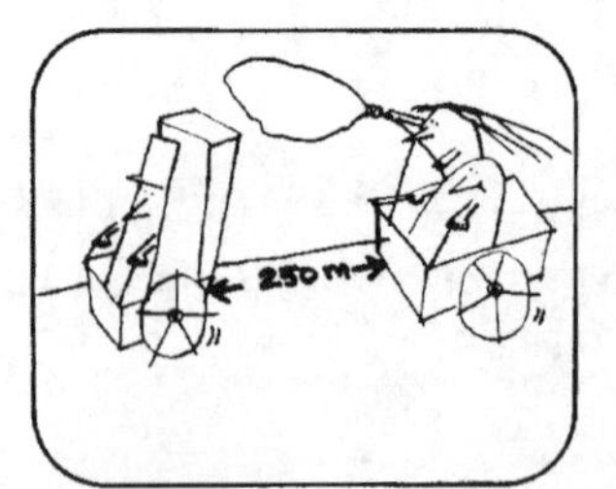

Helen: That's easy. We'd be 250 meters apart 1 minute before the collision.
Helen was correct. Can you explain how she was able to answer so quickly?

Thinking Backward

Although this problem can be solved the hard way by algebra, Helen's insight made such a technique unnecessary. She realized that by running the scene backward in time, the answer could be obtained at once.

这个简单问题提示我们，有些前提或信息对解题是没用的，如题目中的卡车与Bob的小车之间的初始距离就是。

The truck is rolling steadily at 65 kilometers per hour, and Bob is driving a steady 80 kilometers per hour, so his speed relative to the truck is 15 kilometers or 15, 000 meters per hour. This equals 250 meters per minute. Therefore, 1 minute before the crash, the car will be 250 meters behind the truck.

We know that when Bob posed the problem the car was 1.5 kilometers behind the truck. But this information is unnecessary in solving the problem. The answer is the same regardless of the initial distance between the two cars!

There are two classic brain teasers that are both solved by the same time-reversed insight.

1. Two spaceships are moving straight toward each other on a collision course. One ship is going 8 kilometers a minute, the other, 12 kilometers a minute. Assume they are exactly 5,000 kilometers apart. What will be the distance between them 1 minute before they crash?

Here again the distance they are apart at the start is irrelevant to the problem. It misleads many people into thinking that the problem must be solved by considering the initial positions of the spaceships, then moving forward in time. The simple solution, of course, is the realization that the two ships approach each other at a speed of 20 kilometers a minute, so 1 minute before they crash they must be 20 kilometers apart.

2. A molecular biologist developed a strange spore that splits into three spores every hour, each new spore the same size as the original. The three spores in turn, an hour later, each divide into three more, and this process continues indefinitely.

The biologist put a single spore in a container at noon one day. At midnight the container becomes exactly filled. At what time does the container become one-third filled?

The aha! solution, as before, is think backward. Clearly it was one-third filled at 11 o'clock, one hour before midnight.

Now let's test your aha! ability with a new and delightful variation of the last problem. The conditions are exactly the same as before except that the biologist put *three* spores in the same empty container (instead of one)at noon. At what time will the container become filled? The answer appears at the end of the book.

Mysterious Merchandise

神秘的商品

一组数码可以表示一个数，也可以表示一些数码，这种双关意义引出许多游戏。

When they arrived at Helen's house she handed her father a package.
Helen: Here's what you wanted from the hardware store, Daddy.
Mr. Browne: Thanks daughter, how much did they cost?

Helen: Five hundred cost me three dollars.
Mr. Browne: Three dollars? That means they're a dollar a piece.
Helen: That's right Daddy.
What on earth did Helen buy?

Cost Per Numeral

The aha! Here is to realize that "500" can be interpreted in two ways: as a number, or as three numerals. If one numeral costs one dollar, then three numerals will cost three dollars. Helen had purchased three house numbers.

From this problem, you learn that the given information in a stated problem should be analyzed carefully in seeking the problem's solution.

The Unlisted Phone Number

未列入电话簿的电话号码

Bob: By the way Helen, you haven't given me the unlisted phone number of your new house yet.

这个问题是信息论基本概念信息或bit的基础。

Helen: We're not really supposed to give it out, but I will answer 24 yes or no questions about it for you.

Bob: But Helen, there are almost ten million possible phone numbers. How could I ever guess the number in just 24 questions?
Helen: Well, stop and think, Bob. I know you can do it.

It wasn't long before Bob thought of a simple method. It will positively determine anyone's 7-digit phone number in 24 questions or less. If you can figure it out you can try it on your friends.

从数学上讲，可以通过不同进位制来表示，而十进位制最常用。

Binary Sorting

Bob realized that the most efficient way of identifying a specific member of a set, by asking yes or no questions, is as follows: If a set contains an even number of members,

其次常用的是二进位制，熟悉它们之间的转换使许多问题迎刃而解。其他的问题还有称球问题，即用最少次把一堆球中的坏球找出来。

we divide it into equal parts, each containing the same number of elements. If the set contains an odd number of members, we divide it into two parts that are as close to being equal in number as possible. We then ask which of these two parts contains the member we are seeking. After the answer is given, we take the designated part and repeat the same procedure. Eventually only one member of the original set remains. It will be the one we are trying to identify.

One question obviously indentifies specifically a member of a 2-element set. Two questions suffice for a set of 4 elements, three questions for a set of 8 elements, four for a set of 16, and in general, n questions identify a chosen element in a set of 2^n elements.

In our telephone number problem, 24 questions are sufficient for guessing any number no greater than 2^{24} = 16,777,216. This is larger than 9,999,999, the largest possible phone number when the number's seven digits are written as a single number. Twenty-three questions are not enough because 2^{23} = 8,388,608, which is smaller than some phone numbers.

Bob's first question, therefore, is: "Is the number greater than 5 million?" The answer immediately cuts the possibilities in half. Continuing in this manner, he is sure to zero in on the correct phone number in 24 or fewer questions.

Most people find it hard to believe that as few as 24 questions will identify any number from 1 through more than 16 million. This is because they do not realize how rapidly the numbers in a doubling sequence in crease. It is this rapid increase that explains why it is usually easy, by yes-and-no questions, to guess what a person is thinking of even when he is allowed to think of any existing object whatever. If you are skillful in your binary divisions (for example, asking such questions as "Is it living or nonliving?", "Is it animal or vegetable?", and so on), it is often possible in 20 questions or less to guess that someone is thinking, say, of the crown on the Statue of Liberty!

The procedure we described for guessing a phone number in 24 questions is one that computer scientists call a "binary sorting" algorithm. A clever mind-reading trick based on binary sorting uses the six cards shown in Figure 1. Hand a set of these cards to someone and ask the person to think of any number from 1 through 63, then to give you each card that bears the chosen number. You can immediately identify the number.

The secret is simply to add the first numbers on each card given to you. The sum will be the chosen number.

1 Binary mind-reading cards

1	3	5	7	9	11	13	15
17	19	21	23	25	27	29	31
33	35	37	39	41	43	45	47
49	51	53	55	57	59	61	63

2	3	6	7	10	11	14	15
18	19	22	23	26	27	30	31
34	35	38	39	42	43	46	47
50	51	54	55	58	59	62	63

4	5	6	7	12	13	14	15
20	21	22	23	28	29	30	31
36	37	38	39	44	45	46	47
52	53	54	55	60	61	62	63

8	9	10	11	12	13	14	15
24	25	26	27	28	29	30	31
40	41	42	43	44	45	46	47
56	57	58	59	60	61	62	63

16	17	18	19	20	21	22	23
24	25	26	27	28	29	30	31
48	49	50	51	52	53	54	55
56	57	58	59	60	61	62	63

32	33	34	35	36	37	38	39
40	41	42	43	44	45	46	47
48	49	50	51	52	53	54	55
56	57	58	59	60	61	62	63

The cards are constructed by a system that is easily explained by writing the numbers from 1 through 63 in binary notation as shown in Figure 2. The numbers at the left are in decimal form. Each has to its right the same number in binary. The six numbers at the top of the chart are the powers of 2 that are used in forming the binary numbers. The mind-reading card with 1 as its first number bears all the numbers (in decimal form) indicated by a 1 in the last column on the right. The card

2

DECIMAL NUMBERS ↓	BINARY NUMBERS					
	2^5	2^4	2^3	2^2	2^1	2^0
0						0
1						1
2					1	0
3					1	1
4				1	0	0
5				1	0	1
6				1	1	0
7				1	1	1
8			1	0	0	0
9			1	0	0	1
10			1	0	1	0
11			1	0	1	1
12			1	1	0	0
13			1	1	0	1
14			1	1	1	0
15			1	1	1	1
16		1	0	0	0	0
17		1	0	0	0	1
18		1	0	0	1	0
19		1	0	0	1	1
20		1	0	1	0	0
21		1	0	1	0	1
22		1	0	1	1	0
23		1	0	1	1	1
24		1	1	0	0	0
25		1	1	0	0	1
26		1	1	0	1	0
27		1	1	0	1	1
28		1	1	1	0	0
29		1	1	1	0	1
30		1	1	1	1	0
31		1	1	1	1	1
32	1	0	0	0	0	0
33	1	0	0	0	0	1
34	1	0	0	0	1	0
35	1	0	0	0	1	1
36	1	0	0	1	0	0
37	1	0	0	1	0	1
38	1	0	0	1	1	0
39	1	0	0	1	1	1
40	1	0	1	0	0	0
41	1	0	1	0	0	1
42	1	0	1	0	1	0
43	1	0	1	0	1	1
44	1	0	1	1	0	0
45	1	0	1	1	0	1
46	1	0	1	1	1	0
47	1	0	1	1	1	1
48	1	1	0	0	0	0
49	1	1	0	0	0	1
50	1	1	0	0	1	0
51	1	1	0	0	1	1
52	1	1	0	1	0	0
53	1	1	0	1	0	1
54	1	1	0	1	1	0
55	1	1	0	1	1	1
56	1	1	1	0	0	0
57	1	1	1	0	0	1
58	1	1	1	0	1	0
59	1	1	1	0	1	1
60	1	1	1	1	0	0
61	1	1	1	1	0	1
62	1	1	1	1	1	0
63	1	1	1	1	1	1

with 2 as its top number bears all the numbers indicated by a 1 in the second column from the right, and similarly for the other four cards.

These mind-reading cards are easily generalized to notations based on powers of numbers other than 2. Figure 3 shows how to construct a set of cards based on ternary notation. In this case each ternary number may contain 0, 1 or 2. When a 1 appears in a column, we put the corresponding decimal number down once on the card represented by that column. When a 2 appears, we put the number down twice.

3

DECIMAL NUMBERS ↓	TERNARY NUMBERS		
	3^2	3^1	3^0
1			1
2			2
3		1	0
4		1	1
5		1	2
6		2	0
7		2	1
8		2	2
9	1	0	0
10	1	0	1
11	1	0	2
12	1	1	0
13	1	1	1
14	1	1	2
15	1	2	0
16	1	2	1
17	1	2	2
18	2	0	0
19	2	0	1
20	2	0	2
21	2	1	0
22	2	1	1
23	2	1	2
24	2	2	0
25	2	2	1
26	2	2	2

Figure 4 shows a set of three mind-reading cards that identify a chosen number from 1 through 26. Now, however, you ask the person to tell you, each time you are handed a card, whether he/she saw the chosen number once or twice on the card. If he/she saw it twice, you must double the card's top number as you add the key numbers.

同样道理，有时三进位制更有用！

You might wish to extend this system to six cards. As we have seen, the six binary cards identify numbers from 1 through 63. Six ternary cards identify numbers from 1 through 728. It is easy to see how to generalize to bases higher than 3. For example, a set of cards based on powers of 4 have some numbers repeated twice on a card, and some repeated three times. If three times, you must triple the top number before you add.

4

1	14 - 14
2 - 2	16
4	17 - 17
5 - 5	19
7	20 - 20
8 - 8	22
10	23 - 23
11 - 11	25
13	26 - 26

3	15 - 15
4	16 - 16
5	17 - 17
6 - 6	21
7 - 7	22
8 - 8	23
12	24 - 24
13	25 - 25
14	26 - 26

9	18 - 18
10	19 - 19
11	20 - 20
12	21 - 21
13	22 - 22
14	23 - 23
15	24 - 24
16	25 - 25
17	26 - 26

The ternary cards illustrate the fact that "ternary sorting" is more powerful in some ways than binary sorting. If we keep dividing a set into three parts, instead of two, and are told each time which part contains a chosen element, the element can be guessed with fewer questions. Of course, the questions are no longer of a "yes-no" type.

The power of ternary sorting is nicely illustrated by the following card trick. It uses $3^3 = 27$ playing cards. Someone looks through this packet thinks of any card. The magician takes the packet and deals the cards face up into three piles. The person who thought of a card must then say in which pile his card appears.

The piles are assembled by the magician into one packet and again dealt into three face-up heaps. The spectator points to the pile that contains his/her card, the magician assembles the piles, and deals them for a third and last time into three heaps. After being told which heap contains the chosen card, he/she assembles the piles and places the packet face down on the table. The spectator names his/her chosen card. The magician turns over the top card of the packet; and it is the selected card. The trick can be repeated many times, and it never fails to work.

The secret is simple. Each time the magician picks up the three piles he sees that the pile containing the chosen card goes on top when the packet is held face down. This automatically sorts the selected card to the top.

It is not hard to see why it works. The principle is exactly the same as in guessing a telephone number, except that instead of dividing the set of elements in half each time, they are divided into thirds. After the first pick-up, the card must be in the top nine. After the second pick-up, it must be among the top three. After the third pick-up it must be the top card. If you run through the procedure, with the selected card turned face up, you will be able to follow its progress as it moves upward, in three stages, to the top. The sorting of elements by computers, using procedures such as this, plays a major role in modern information retrieval theory.

Hapless Hat

倒霉的帽子

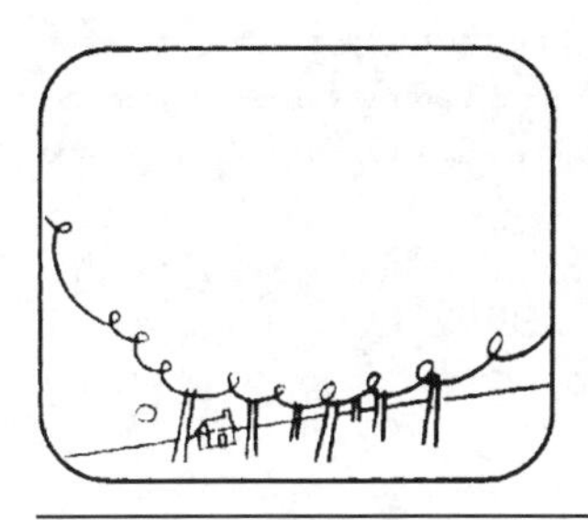

Bob and Helen decided to spend their vacation in the Maine woods where Uncle Henry lived in a cabin.

To get to the cabin they had to rent a canoe and paddle up a river.

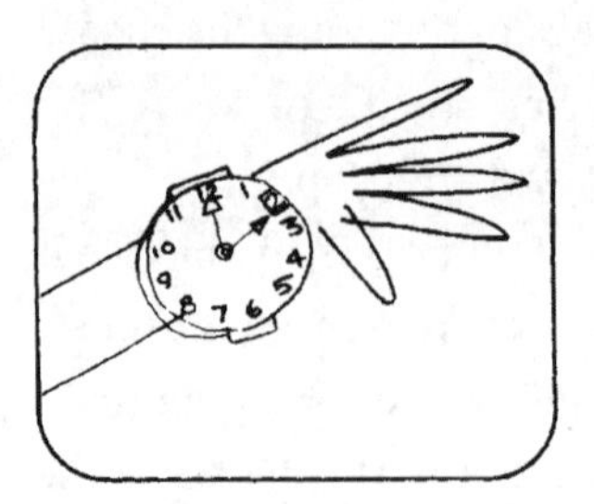

Bob took the bow position and Helen paddled stern. At two o'clock she took off her straw hat and put it behind her on the stern.

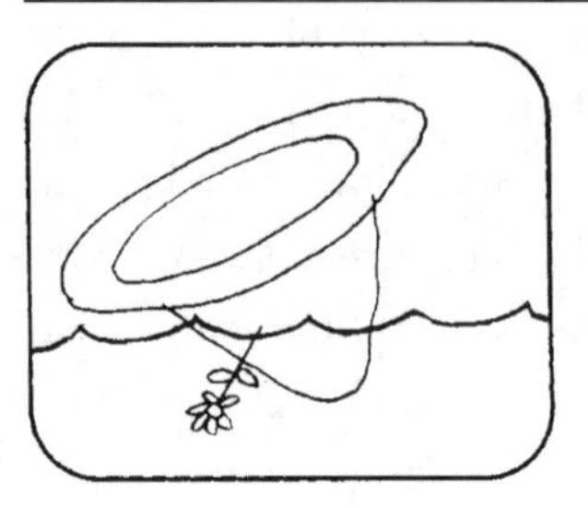

Then a gust of wind blew it off without Helen or Bob noticing it at the time.

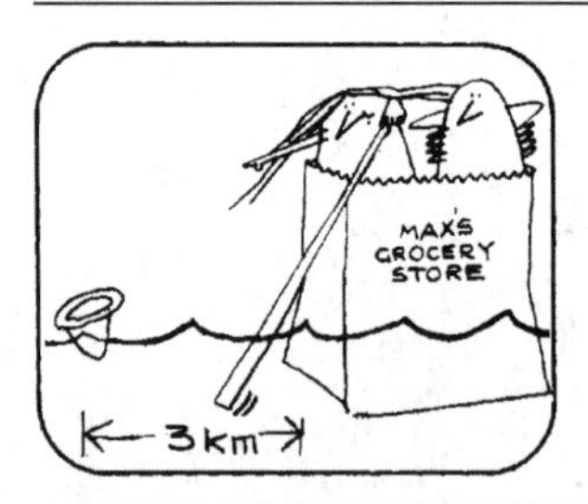

Only after they had paddled 3 kilometers upstream from the hat did Helen suddenly shout out.

Helen: Wait! Stop the canoe. I've lost my beautiful hat.

They turned the canoe around and paddled downstream until they came to the hat.

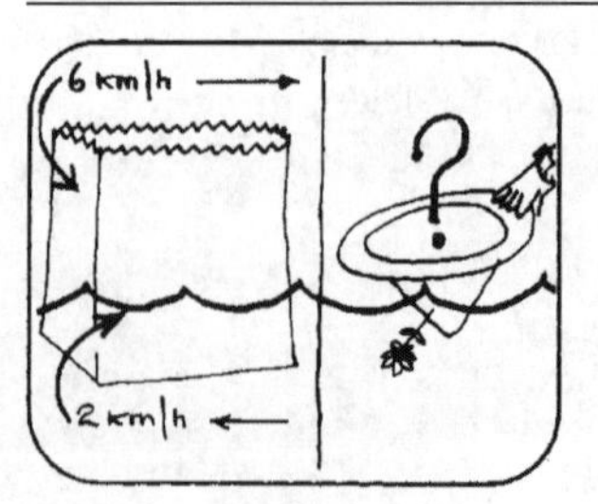

Assuming that the canoe's speed through the water was always 6 kilometers per hour, and that the river flowed at a steady 2 kilometers per hour, at what time did Helen retrieve her hat?

Did you get the aha! that makes the solution easy? Believe it or not the speed of the water has the same effect on both canoe and hat, and can be completely ignored.

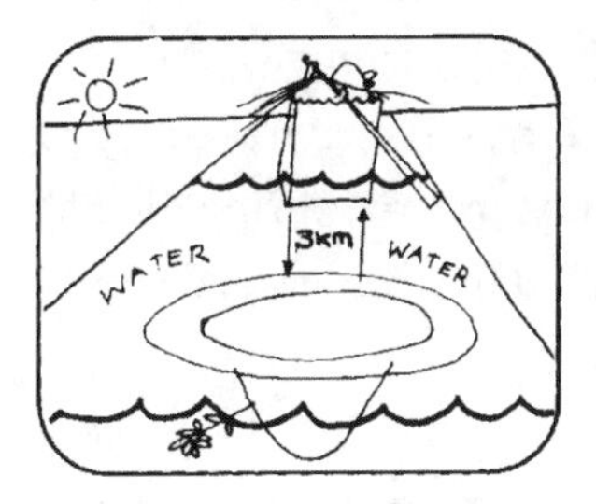

So, with respect to the water, the canoe travels 3 kilometers away from the hat and then 3 kilometers back. A total of 6 kilometers.

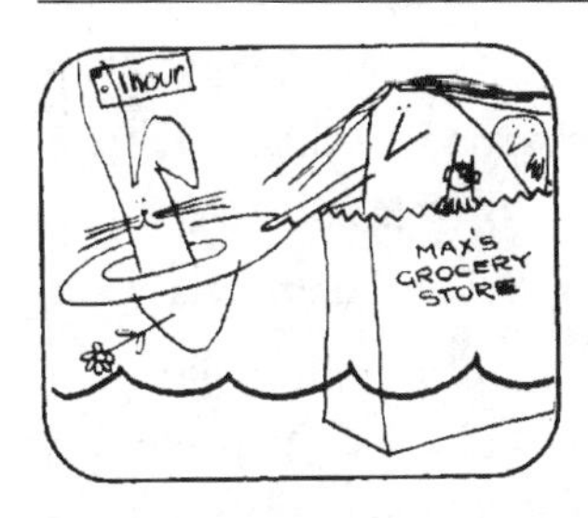

And because the canoe goes 6 kilometers per hour, the up and down trip will take just an hour. The trip makes it 3 o'clock when Helen picks up her hat, doesn't it?

Relative Speeds

Helen and Bob made a roundtrip upstream from a hat and back to it. The current had no affect on their travel time because the hat was carried along with the current. Here is a good variation in which the roundtrip is made not from an object moving with the current but a fixed object on the shore.

速度问题是算术课极好的课外练习，容易通过代数方法解决。

Assume there is no current in the river. Bob and Helen row upstream for 3 kilometers from a boathouse on the shore, then turn around and row back to it. The travel time for the roundtrip is 20 minutes.

Now suppose that the river is flowing downstream at 2 kilometers per hour, as in the previous problem. If they row 3 kilometers upstream, then back to the boathouse, will the trip's total time be longer or shorter than 20 minutes?

One is tempted to say that it will still be 20 minutes

because the current will slow down the canoe while it goes upstream by the same amount as it speeds up the canoe while it goes downstream.

This is not correct. Why?

The insight that answers this question is as follows. The trip upstream for 3 kilometers will take *longer* than the trip downstream for 3 kilometers. Therefore the canoe will be slowed down by the current for a longer period of time than the time during which it is speeded up by the current. Consequently, the roundtrip will take longer. This can easily be verified by setting up algebraic equations.

The same insight applies to airplanes that travel with and against a wind. If a plane takes a certain time to go from A to B and back to A when there is no wind, it is sure to take longer to make the same roundtrip if there is a steady wind blowing from A to B, or from B to A.

Another good problem involving motion relative to a fixed object on land is the following: A girl gets on the last car of a train. She can't find a seat, so she leaves her heavy suitcase in the vestibule just as she is passing the Flat Foot Shoe Factory. She walks forward through the train at a steady speed until 5 minutes later she reaches the front car. Having found no seat, she turns around and walks back at the same constant rate until she comes to her suitcase. At that moment she is passing the Flat Head Wig Factory which is just 5 kilometers from the Flat Foot Shoe Factory. How fast is the train going?

As in the first problem, a simple aha! leads at once to the solution. It is not necessary to know how fast the girl walks or how far she walks. If it takes her 10 minutes to make the roundtrip up and down the aisles of the train, the suitcase will have traveled 5 kilometers in the 10 minutes. Therefore, the train has a speed of half a kilometer per minute or 30 kilometers per hour.

Here's a little-known speed puzzle that confuses even good mathematicians. A boy and girl ran a 100-meter race. The girl crossed the finish line when the boy had gone 95 meters, so she won the race by 5 meters.

When they raced a second time, the girl wanted to make the contest more even so she handicapped herself by starting 5 meters behind the start line. If the two ran at the same constant speed as before, who won the second race?

If you think it was a tie, you'll have to think again and search for an aha! (Hint: At what spot along the track will the boy and girl be neck and neck?)

An amusing quickie concerns an intoxicated ladybug at one end of a meter stick. She wants to crawl to the other end. Every second she goes 3 centimeters forward and 2 centimeters backward. How long will it take the looped lady to reach the end of the stick? (The answer is *not* 100 seconds.)

Money Matters

钱币问题

Just before they reached Uncle Henry's place Helen gave this quickie to Bob.

Helen: Which is worth more? A piggy bank filled with five-dollar gold pieces, or the same bank filled with ten-dollar gold pieces?

Bob was stumped for a while but got the right answer eventually. Then he gave this one to Helen in return.

Bob: A Scotsman had 44 single dollar bills and ten pockets. How can he distribute the money so that each pocket contains a different number of bills?

Pigeon Hole Proof

A piggy-bank filled with five-dollar gold pieces contains the same amount of gold as a piggy-bank filled with ten-dollar gold pieces, therefore the gold in each is worth the same. You might think that small coins would pack

鸽洞原理是数论中最重要的原理之一。

如果鸽子数大于鸽洞数，则至少有一个鸽洞中有两只或两只以上鸽子。当前在丢番图逼近等方面有重要应用。

a bank with greater density than large coins, but this is not the case. If you fill a bucket with tiny pebbles, the proportion of air space to the volume of the bucket is the same as when you fill the bucket with big pebbles.

The problem of the Scotsman with 44 single dollar bills and ten pockets is even trickier. Let's see what happens when we put the smallest amounts possible into the pockets. The first pocket contains a zero number of bills, the second contains one bill, the third contains two bills, and so on until the tenth pocket contains nine bills. But 1 + 2 + 3 + 4 + 5 + 6 + 7 + 8 + 9 = 45, so we have already gone beyond the 44 available bills. Obviously there is no way to cut down on the number of bills in any pocket without duplicating two numbers for a pair of pockets.

Mathematicians call this type of proof a "pigeon hole" proof. Here is an amusing example of another problem solved by the same technique. Suppose that a town contains no more than 200,000 people. Do two inhabitants of the town have exactly the same number of hairs on their head?

At first thought you may consider this unlikely. But let's see what happens when we apply the pigeon hole analysis. The number of hairs on one person's head does not exceed 100,000. If there are no matching heads, then one person could be bald, another could have one hair, another could have two, and so on. But as soon as we pass 100,000 people with distinct numbers of hairs on their heads, we are forced to duplicate. The 100,001th person is certain to have a head of hair that matches someone among the 100,000. Since the town has a population of about 200,000, it is sure to have not just two people with matching heads, but about 100,000 people with matching heads!

Uncle Henry's Clock

亨利叔叔的钟

Helen had just answered Bob's quickie when they arrived at Uncle Henry's. His cabin, which he had built himself, had no electricity, phone, TV or radio.

The first thing Uncle Henry said was.
Henry: What time is it?

Helen: Sorry Uncle, we lost our watches on our way up here. Don't you have a clock?
Henry: Yep. But, dang it, I forgot to wind it last night. You two stay here while I walk to the village to check the time and pick up some vittles.

Uncle Henry walked to town and spent about half an hour at the grocery store.

And when he got home, the first thing he did was to set his clock.

Helen: Are you sure that's the correct time? It can't be unless you know how far you walked and how fast you walked.

Henry: Nope, Helen. I don't know none of them things. All I know is when I go to town and back the same way, and go the same speed, I can always set my clock right.

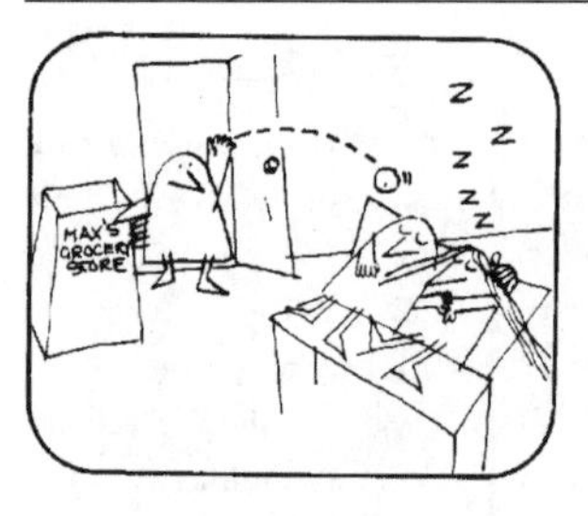

Supposing that Uncle Henry wound his clock before he left, and that the grocery store's clock is accurate, how did he know the exact time when he got home?

Setting the Clock

钟表问题是一个有趣的算术问题，但需要一定的智力。

The aha! insight that leads to a solution of this problem is the realization that Uncle Henry can wind his stopped clock before he leaves, and use it to determine the total time that elapses between leaving home and returning. He cannot, of course, set the clock correctly after he winds it, because he does not yet know the correct time. He does, however, note the time on the clock before he leaves.

When he gets back, the clock tells him how long it took for him to walk to town, spend time at the grocery store, then walk back. Since there is a clock at the store, he has no trouble determining the time he spent at the store. He subtracts this from the total time he was away from his house (as measured by the clock at home) to get the time he spent walking to and from town. Because he always walks the same way, at the same constant rate, half his walking time is the time it takes him to walk home. He then adds this to the time on the store clock when he left, and that gives him the correct time of his arrival home. Since he sees exactly when he arrived home, he is able to set his house clock correctly.

Here is a tricky clock question that nine out of ten people answer incorrectly. How many times does the minute hand pass the hour hand between 12 o'clock noon and 12 midnight? Most people say 11 times, but the correct answer is 10! If you don't believe it, try moving the hands of your watch to convince yourself it is true.

This somewhat surprising fact is involved in the solution of a problem that seems at first to be unsolvable without writing algebraic equations. A clock has a sweep second hand. At 12 noon all three hands coincide. Is there another time, before it is 12 o'clock again, when all three hands are exactly together?

Let us first determine at how many spots the hour and minute hand coincide. You might think that they coincide at 12 spots, but, as we have seen, this occurs only 10 times between 12 noon and 12 midnight. The coincidence of the hands at 12 makes a total of 11 different spots at which the two hands coincide. By the same reasoning, the second hand and minute hand coincide at 59 different spots. Thus, the coincidences for the hour and minute hand are separated by 11 equal time periods, and the coincidences for the minute and second hand are separated by 59 equal time periods.

Let us call the number of intervals between the first coincidences A, and the number of intervals between the second coincidences B. If A and B have a common factor k, there will be k spots where the two coincidences will occur simultaneously. But 11 and 59 have no common factor. Therefore, there cannot be a spot between 12 noon and 12 midnight when both coincidences occur at the same time. In other words, the three hands are exactly together only at 12 o'clock.

Now for two quickie clock questions that catch most of your friends. A clock takes 5 seconds to strike 6 o'clock. How many seconds will it take to strike 12 o'clock?

Suppose Uncle Henry was so tired that he went to bed at 9 o'clock with plans to sleep until 10 the next morning. He set his alarm clock for 10, and fell asleep 20 minutes later. How long did he sleep before the clock woke him up?

Both quickies are answered at the end of the book.

Spirits of 1776

1776 年的精神

最后一节是数论最重要的内容，即模算术。模的思想来源于高斯（Johann Carl Friedrich Gauss）。

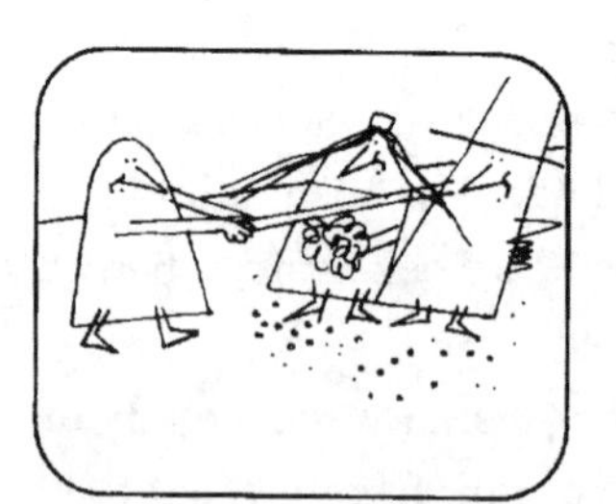

On the last day of their visit, Bob and Helen told Uncle Henry they had decided to get married.

Uncle Henry: Wonderful, my dears. This calls for a celebration.

Uncle Henry then produced five bottles of wine that he had been saving for a special occasion. But nobody could agree on which bottle to open.

Uncle Henry: I know. Let's put the bottles in a row. Then I'll count back and forth according to my lucky system. Here's how it works. One, two, three, four, five, ...

Uncle Henry: Six, seven, eight, nine,...

Uncle Henry: Ten, eleven, twelve, thirteen,... Get the idea?

Bob: Yes, I do Uncle, but how high are you going to count up to?

Uncle Henry: Ain't this the bicentennial year, 1976? Let's count to 1976.

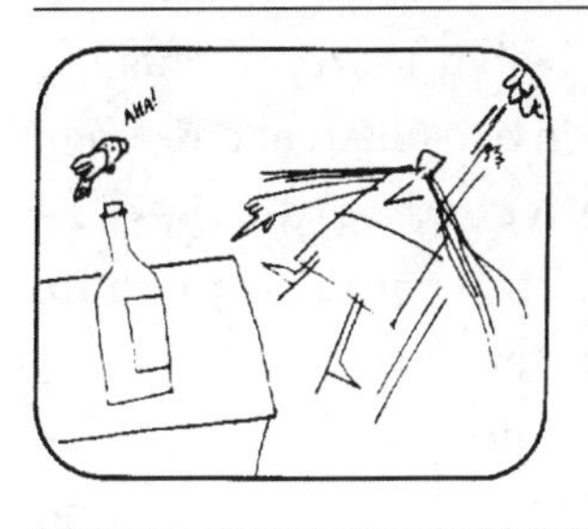

Helen: (groan) Oh dear, Uncle Henry, that will take forever. Hmm. Wait a minute. You don't have to count. I can tell you right now where the count will end.

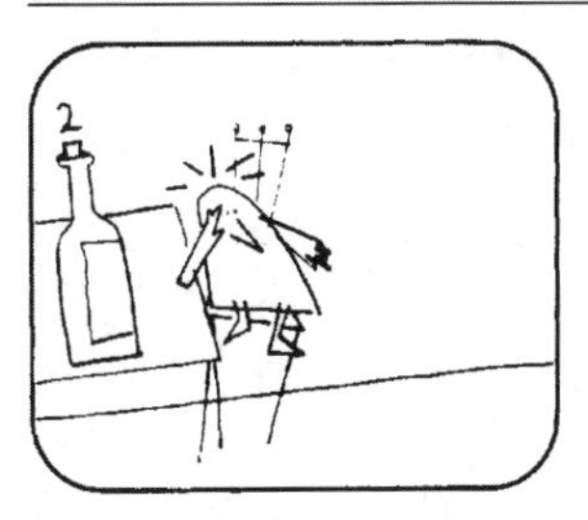

Helen: It'll end up on the second bottle. I've just figured it out. Uncle Henry didn't believe her and had to count the bottles himself. Fifteen minutes later he ended his count on the second bottle.

Uncle Henry: Heavens to Betsy. How did you know, Helen?

See if you can figure out an easy way to tell where the count will end, no matter how big the number counted. You might want to try some variations on your friends.

Modulo Arithmetic

Helen's insight, which avoided tedious counting of the bottles from 1 to 1,976, was the realization that the question could be answered quickly by applying what is called modulo or clock arithmetic.

A clock models a finite arithmetic of just 12 numbers. Actually, 12 corresponds to 0 in a modulo arithmetic based on 12. Suppose it is 12 o'clock, and you wish to know what time the clock will show 100 hours later. This can be calculated merely by dividing 100 by 12 and noting the remainder. The remainder, 4, tells us that the time will be 4 o'clock. Only the remainder concerns us. The number 100 is said to be equal to 4 (modulo 12), meaning no more than that 4 is the remainder when 100 is divided by 12.

Do you see how Uncle Henry's method of counting is equivalent to clock arithmetic? The only difference is that each of the three middle bottles represents *two* numbers because it is counted in two different directions. The count of 8 ends on the second bottle, then the counting cycle begins again. The counting procedure therefore models a modulo 8 arithmetic.

Helen had only to determine the value of 1,976 (modulo 8). In other words, she divided 1,976 by 8 and obtained a remainder of 0. In modulo 8 arithmetic, 8 = 0 (modulo 8), therefore the count of 0,976 must end on the second bottle from the end where the counting started.

Suppose you wanted to know where Uncle Henry's count would end if he counted to a large number such as 12,345,678,987,654,321. Is it necessary to divide this entire number by 8? No, not if you have another aha! Since 1,000 = 0 (modulo 8) you need only divide the last three digits, 321, by 8 to get a remainder of 1. This tells you that 12,345,678,987,654,321 = 1 (modulo 8), so the count will end on the first bottle.

By changing the number of bottles, you produce models of finite arithmetics with other even modulos. If the bottles are counted in the usual manner, from left to right only, then you can model a finite arithmetic with any modulo, odd or even.

A famous problem involving the counting of objects in a cyclical manner is called the Josephus problem because it dates back to an ancient Roman story that involves a man named Josephus. There is a large literature on the problem and its many variants. Here is a new version you will find amusing.

Once upon a time, a rich king had a beautiful daughter named Josephine. Hundreds of young men wanted to marry her. She finally eliminated all of her suitors except the ten she liked best.

Several months went by and the king became annoyed because Josephine couldn't make up her mind. "My dear," he said, "next month you'll be 17. As you know, it's the custom of all princesses to marry before that age."

"But father," she replied, "I'm still not sure that I like George the best."

"In that case, my pretty one, we'll have to settle the matter today by our secret ritual."

The king then explained to his daughter how the ancient ritual worked. "The ten men," he said ,"will stand in a circle. You can pick any man you like and call him 1. Then you must start counting clockwise around the circle of men until you reach 17—your age. The 17th man must drop out of the circle. We'll send him back home with a consolation gift of 100 gold pieces."

"After he is gone, you must count again from 1 to 17, this time starting your count on the next man after the one who dropped out. When you reach 17, the 17th man will be eliminated as before. Continue doing this, always counting the men that remain, until only one is left. He'll be the man you must marry."

Josephine frowned and said: "I'm not sure I understand, father. Do you mind if I practice it once using 10 gold pieces?"

The king agreed. Josephine put the ten gold pieces in a circle and counted around, removing every 17th piece until only one remained. The king watched, and saw that his daughter understood the secret ritual perfectly.

The ten suitors were then ordered into the throne room. They formed a circle around Josephine.

Without hesitation she started her count on Percival, and counted rapidly until every man was eliminated except George, the man she had privately decided she wanted to marry.

What insight did Josephine have that made it easy for her to begin a count that she knew would end by selecting George?

Here is how Josephine managed it. When she made her practice count with the gold pieces she remembered that the single piece that remained was number three from the spot where she began the count. So when she started to count the men, she began the count at a spot that would give George the count of 3.

An interesting generalization of the Josephus problem can be modeled with the thirteen spades of a deck of playing cards. Can these cards be arranged into a sequence so that you can perform the following Josephus count?

The count begins with the packet of 13 cards held face down in one hand. Call the top card 1, turn it over, and it is the ace of spades. Deal the ace to the table. Then count 1, 2, placing the first card beneath the packet. The second card is turned over and it is the two of spades. Deal it to the table. Now count 1, 2, 3, putting the first two cards beneath the packet, and turning the third one. It is the three of spades. Deal it to the table. Continue in this way, transferring cards one at a time from top to bottom (which is the equivalent of a Josephus count around a circle) until you have correctly turned over and dealt the thirteen spades in consecutive order from ace to king.

Here is how the cards are arranged, from top down, to make such a count possible: A, 8, 2, 5, 10, 3, Q, J, 9, 4, 7, 6, K.

You might suppose that it took someone many hours of trial and error to devise such a clever arrangement. Actually, there is a very simple algorithm (procedure) for obtaining such sequences. Many magicians, working on counting tricks of this sort, have indeed wasted vast amounts of time before they had the aha! that made the task trivial. See if you can think of it before you read the solution at the end of the book.

Chapter 4

Logic aha!

逻　　辑

Puzzles about reasoning

有一位著名的数学家说：逻辑是数学的卫生。数学不能简单地还原为逻辑，但是逻辑隐含在每一步数学推理之中。逻辑混乱与矛盾会造成数学错误甚至危机。因此逻辑思维是每个学习数学的人必须要好好掌握的。

In this section we are not concerned with formal logic, but with problems that can be solved by reasoning, without any special expertise in mathematics. Some of the short puzzles are close to riddles in the sense that they contain deliberately misleading statements, or the answers hinge on word play, but most of them are puzzles that play fair with the reader.

There is a general way in which logic puzzles of this sort are related to mathematics. All mathematical problems are solved by reasoning within a deductive system in which basic laws of logic are embedded. Although you need not know formal logic to work on any of the problems in this section, the informal reasoning that solves them is essentially like the reasoning that mathematicians and scientists use when confronted with a perplexing question.

By "perplexing" we mean a problem of such a nature that one does not know how to go about solving it. Naturally, if there is a known procedure—for example, the technique for cracking a quadratic equation—everything is cut and dried and there is no real perplexity. One simply applies the proper algorithm and grinds out the answer.

The interesting and challenging problems that arise in mathematics and science are those for which the method of solution is not apparent. One must think long and hard about the question, searching the memory for all relevant information, and hope for that moment of aha! insight that suggests a solution. In this general way, the solving of amusing logic puzzles is good training for solving more serious problems.

Several puzzles in this chapter have even closer ties to significant mathematics. For example, "Color Mates" and the problems that follow it lend themselves to a chart method of solution that is very similar to techniques used in formal logic. One of these puzzles introduces an important logic relation called "material implication". In the propositional calculus (a fundamental branch of symbolic logic) implication is symbolized by ⊃. The relation $A \supset B$ means that if A is true, then B must be true. It is one way of interpreting the statement in set theory that all of set A is included in set B.

The word "induction" has two essentially different meanings. Scientific induction is a process by which scientists make observations of particular cases, such as noticing that *some* crows are black, then leap to the universal conclusion that *all* crows are black. The conclusion is never certain. There is always the possibility that at least one unobserved crow is not black.

Mathematical induction, to which you are introduced in the comments on the hat tests in "Dr. Ach's Awards", is an entirely different procedure. Although it, too, leaps from the knowledge of particular cases to knowledge about an infinite sequence of cases, the leap is purely deductive. It is as certain as any proof in mathematics, and an indispensable tool in almost every branch of mathematics.

Most of the puzzles in this section are not as serious or as complicated as the hat problems. Nevertheless, they are sure to sharpen your wits. They will teach you the value of looking carefully for verbal pitfalls in the statement of a problem, and, above all, the value of going out on a limb in considering offbeat possibilities. The more possibilities you consider, however bizarre the more likely it is that the right insight will come. It is one of the secrets of all creative mathematicians.

The Crafty Cabbie

狡猾的司机

无论是在数学还是在日常生活中，直来直去往往不合适或者行不通，因此常常出现"换一种说法"

One day, this lady in New York City hailed a passing taxicab.

On the way to her destination the lady talked so much that the driver got quite annoyed.

——用数学的词汇来说就是等价的说法。但是换一种说法往往出现漏洞或矛盾，逻辑就是揭示这种矛盾的有力武器。最简单情况就是A与非A不能同时成立。

Driver: I'm sorry lady, but I can't hear a word you're saying. I'm deaf as a post, and my hearing aid hasn't worked all day.

When she heard this, the lady stopped yakking. But after she left the cab she suddenly realized that the cabbie had lied to her. How did she know?

Observant Lady

The story of the lady and the cab typifies many situations in both ordinary life and in science. There is a puzzling situation which at first one cannot understand. But if all the relevant factors are carefully considered, suddenly the mind has a flash of insight into a forgotten aspect of the problem that furnishes a key to its solution.

If you cannot answer the crafty cabbie puzzle right away, try to put yourself in the position of the lady, then in your mind act out the entire sequence of events. What is the first thing you say when you get in a cab? The answer, of course, is that you tell the cab driver where you want to go. But if the driver is deaf, how will he know where to take you? The lady suddenly realized, after she had paid the fare, that the cabbie could not be deaf because he drove her to her proper destination.

Logic puzzles based on real life situations are often not well-defined. They frequently require many unstated assumptions, and this problem is no exception. For example, it may have occured to you that if the cabbie saw the lady's face when she told him her destination, he might have read her lips. This is not an irrelevant quibble, but a shrewd observation on your part.

A careful analysis of every aspect of a sequence of events has often led to major insights in the history of science. A beautiful example was the solution to the puzzling question of how worker bees know where to go to obtain a supply of honey that has been discovered by one worker bee that returns to the hive. Karl von Frisch observed that when the scouting bee returns, it engages in a curious kind of "dance" . Could it be that the nature of this dance communicates the destination of the honey source? In a brilliantly designed series of elegant experiments, von Frisch finally proved that this is indeed the case.

If you found the crafty cabbie puzzle amusing, here are two other taxicab problems. A cabbie picked up a customer at the Waldorf Hotel in New York City who wanted to go to Kennedy Airport. The traffic was heavy, and the cab's average speed for the trip was 30 kilometers per hour. The total time of the trip was 80 minutes and the customer was charged accordingly. At Kennedy Airport, the cabbie picked up another passenger who, by coincidence, wanted to be taken to the Waldorf Hotel. The taxi driver returned to the hotel along the same route he had traveled before, with the same average speed. But this time the trip took an hour and twenty minutes. Can you explain why?

It may take a while before it dawns on most people that 80 minutes is the *same* as one hour and twenty minutes! It is an amusing catch puzzle to try on friends.

Another catch problem involving a taxicab goes like this:

You are a taxi driver. Your cab is yellow and black, and has been in use for seven years. One of its windshield wipers is broken, and the carburetor needs adjusting. The tank holds 20 gallons, but at the moment is only three-quarters full. How old is the taxi driver?

This is even a bigger swindle than the previous problem, although logically it is perfectly consistent. You were told at the outset that *you* are the driver. Therefore, the driver is whatever age *you* are!

Color Mates

颜色的搭配

这个问题就是最好的逻辑游戏。它最简单，但需稍微考虑一下才能得出正确结论。

The cabbie next picked up three young couples and took them to a discotheque. One girl was dressed in red, one in green, and one in blue. The boys wore outfits of the same three colors.

When all three couples were dancing, the boy in red danced close to the girl in green and spoke to her.

Frank: Isn't it funny, Mabel? Not one of us is dancing with a partner dressed in the same color.

Given this information, can you deduce the color the partner of the girl in red is wearing?

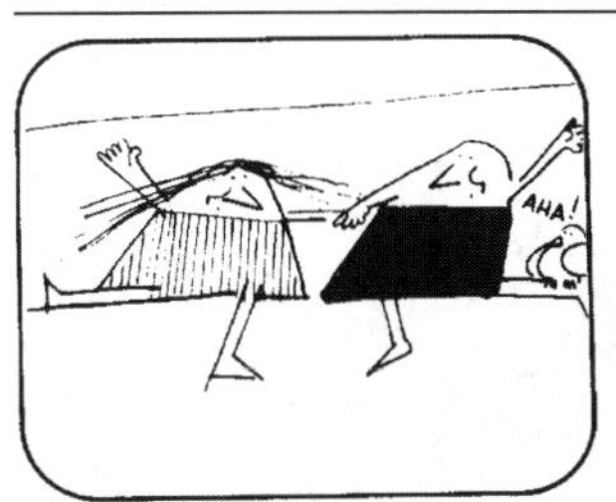

The boy in red must be with the girl dressed in blue. She can't be red because then they would match. And she can't be green because the boy in red spoke to the green girl when she was dancing with someone else.

The same argument shows that the girl in green can't be with either of the boys in red or green. So she must be with someone in blue.

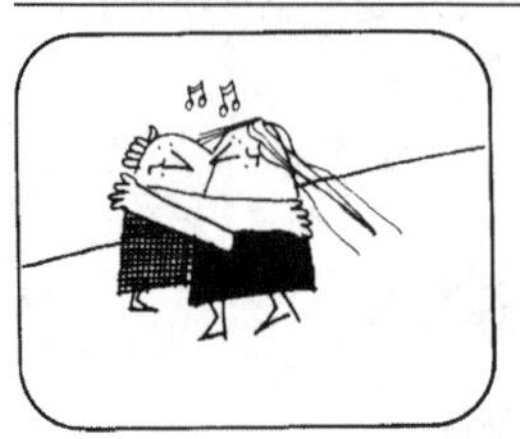

That leaves the girl in red with the boy in green and our problem is solved, isn't it?

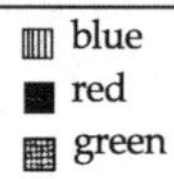

当条件多时，许多人绕不过来，这里教一个比较机械的方法。

Color Opposites

Most people do not find it easy to follow the reasoning in the solution to this problem. One is not likely to have an aha! insight until one fully understands what is being asserted by each statement. A good way to organize this information is to classify it on a square matrix of the type shown below:

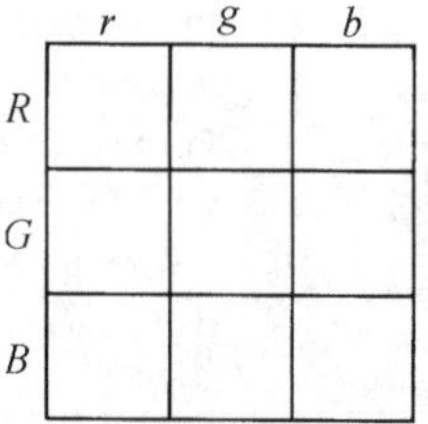

The capital letters on the left side of the matrix stand for the colors of the boys: R = red, G = green, B = blue. The lower case letters at the top stand for the colors of the girls.

We are told that no boy matches a girl in color. We can, therefore, eliminate three possible combinations: *Rr*, *Gg* and *Bb*. This is indicated on the matrix by shading the corresponding three cells:

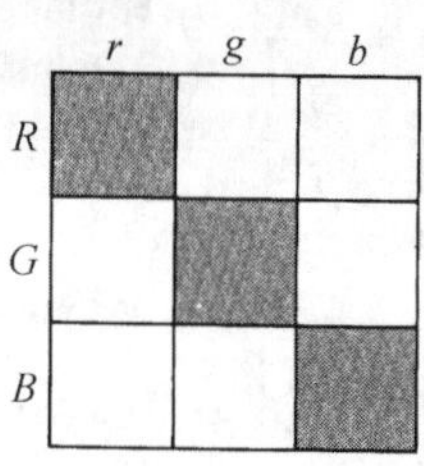

Because the boy in red danced over to the girl in green, we know he is not with a green girl. This allows us to eliminate the *Rg* cell. Now only one cell remains on the *R* row, which proves that the boy in red is with the girl in blue. We indicate this by putting a check in the *Rb* cell. Our chart now looks like this:

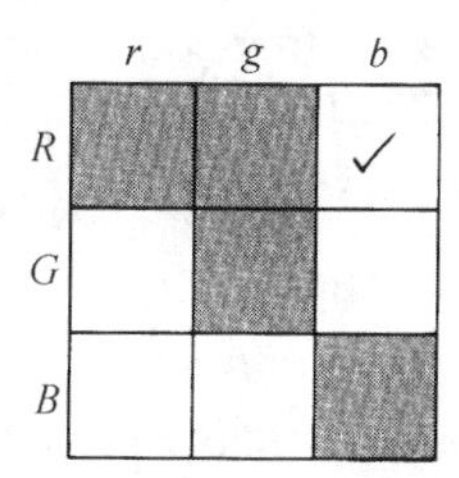

Since we know that the girl in blue is with the boy in red, she cannot be with any other boy. Therefore, we can shade the *Gb* cell. Only the *Gr* cell remains open in the second row. This tells us that the green boy is with the *r* girl, so we put a check in that cell:

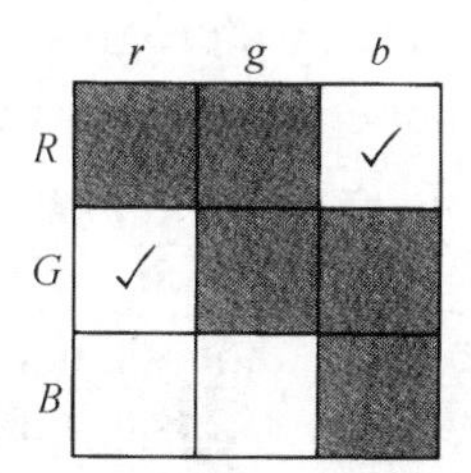

Since the red girl is with the green boy, she cannot be with any other boy, so we shade the *Br* cell. This leaves open only the *Bg* cell, so it gets a check to show that the boy in blue is with the girl in green. Our problem has been solved:

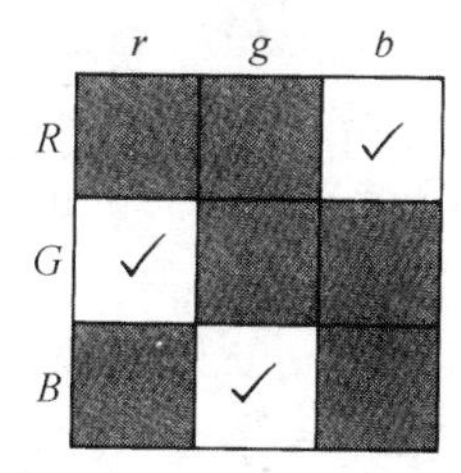

Here is a more difficult logic problem of essentially the same kind. Very few people can solve it without the help of a matrix.

Paul, John and George are three rock stars. One plays

a guitar, one plays the drums, and one plays the piano, but not necessarily, respectively.

The drummer tried to hire the guitarist for a recording session, but was told that the guitarist was out of town doing shows with the pianist.

1. The pianist earns more money than the drummer.

2. George earns less than John.

3. George has never heard of John.

4. What instrument does each of the rock stars play?

See if you can draw the 3-by-3 matrix, and eliminate all impossibilities in the manner previously explained. If you do it properly, you will obtain the following correct answer: Paul plays the guitar, John the drums, and George the piano.

三个摇滚明星问题的正确答案似乎不对，John 应演奏钢琴（piano），George 击（演奏）鼓。否则与第 1、第 2 条矛盾。

Solving logic problems by the use of such charts is very similar to the techniques of solving problems in formal logic by the use of Venn diagrams. In both cases, solutions are obtained by a progressive elimination of impossible combinations of "truth values" until only one combination, the correct one, remains. As Sherlock Holmes, in *The Sign of Four*, once said to Watson: "When you have eliminated the impossible, whatever remains, however improbable, must be the truth."

Here is a problem more challenging than the previous ones. It will introduce you to a fundamental binary relation in formal logic that is known as "implication". It is a statement that has the form: "If... then ..."

Four college girls who share an apartment are listening to an album of music while one of them does her nails, one does her hair, one puts on make-up, and one is reading.

1. Myra isn't doing her nails and she isn't reading.

2. Maud is not putting on make-up and she is not doing her nails.

3. If Myra is not putting on make-up, Mona is not doing her nails.

4. Mary is not reading and she is not doing her nails.

5. Mona is not reading and she is not putting on make-up. What is each girl doing?

You should have no difficulty drawing a 4-by-4 matrix for the four girls and the four tasks. Statements 1, 2, 4 and 5 each eliminated two cells.

Statement 3 is the statement of implication. It asserts that *if* Myra is not putting on make-up, *then* Mona is not doing her nails. Let *A* stand for the "if" clause, and *B* for the "then" clause. The "if-then" binary relation tells us that the truth of *A* cannot be combined with the falsity of *B*, but it tells us nothing about the truth values of *A* and *B* when *A* is false.

Statement 3, therefore, allows for the following three combinations of truth values:

1. Myra is not putting on make-up, and Mona is not doing her nails.

2. Myra is putting on make-up, and Mona is not doing her nails

3. Myra is putting on make-up, and Mona is doing her nails .

After you have eliminated eight "impossible" combinations by shading the eight cells that are ruled out by statements 1, 2, 4 and 5, you will then have to test each of the three possible combinations given by statement 3. Two of them lead to logical contradictions; that is, to two girls doing the same task. Only the combination of "Myra is doing her make-up, and Mona is doing her nails" does not conflict with the information provided by the other statements. The final solution is:

Myra is putting on make-up.
Maud is reading.
Mary is doing her hair.
Mona is doing her nails.

A shorter solution, proposed by Peter Stangl, is to recognize that since statements 1, 2, 4 and 5 show that neither Myra, Maud, nor Mary are doing their nails, therefore Mona must be the girl who is doing her nails. This contradicts the second part of the "if-then" assertion of statement 3, therefore the first part of this assertion must also be false. Consequently, Myra is putting on her make-up, and this leaves Mary as the girl doing her hair.

Logic puzzles of this sort are not hard to invent. You might enjoy trying your skill at designing one yourself. There are many different techniques for solving such problems—algebraic techniques, methods using graph theory, different types of logic diagrams, and so on. Maybe you can also invent a method of your own that is as good or better than the matrix method given here.

Six Sneaky Riddles

六则怪谜语

解答这些谜语的关键是不作错误的假设。

When the music stopped the six friends returned to their table and amused themselves by asking each other riddles. How many can *you* get?

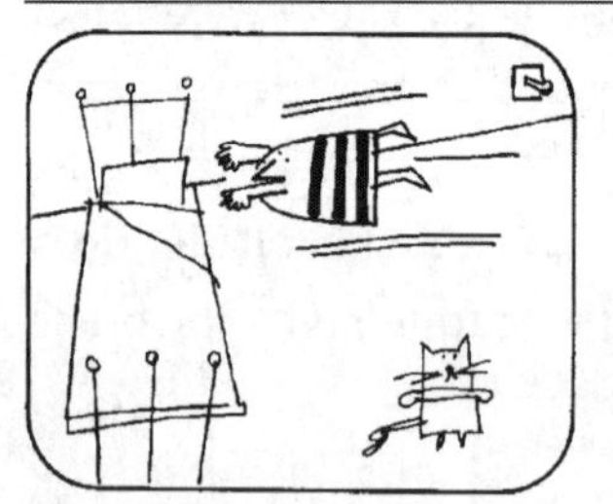

The boy in red asked his first.

Frank: Last week I turned off the light in my bedroom and managed to get to bed before the room was dark. If the bed was 10 feet from the light switch, how did I do it?

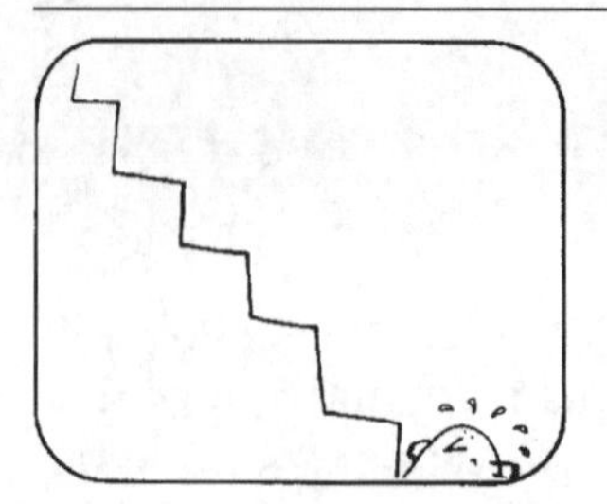

The boy in blue said.

Henry: Whenever my Aunt comes to visit me at the apartment she always gets off five floors too soon and walks up the rest of the way. Can you tell me why?

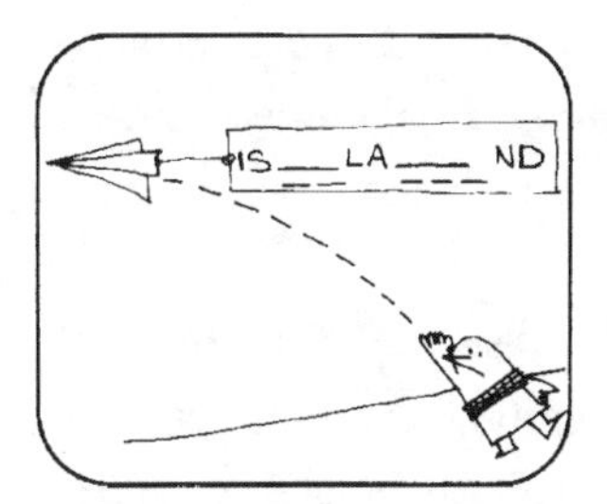

The boy in green said.
Inman: What common word starts with "IS", ends with "ND," and has "LA" in the middle?

The girl in red said.
Jane: One night my uncle was reading an exciting book when his wife turned out the light. Even though the room was pitch dark he went right on reading. How could he do that?

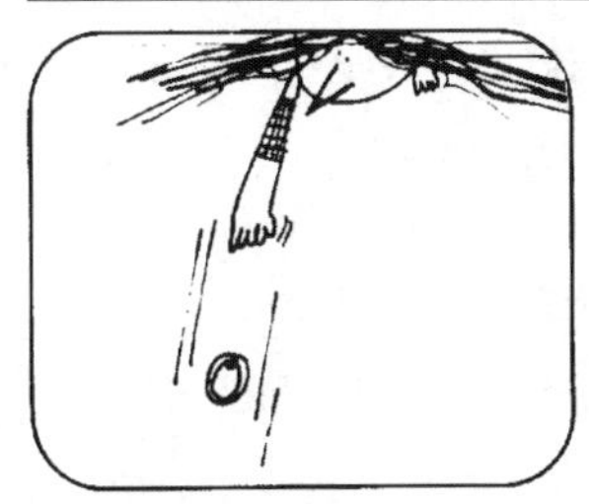

The girl in green said.
Mabel: This morning one of my earrings fell into my coffee.Even though my cup was full the ring didn't get wet. How come?

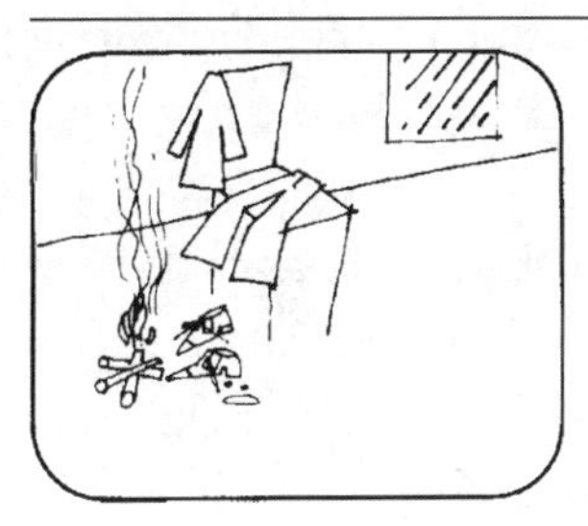

The girl in blue asked the last riddle.
Laura: Yesterday, my father was caught in the rain without hat or umbrella. There was nothing over his head and his clothes got soaked. But not a hair on his head got wet? Why was that?

Sneaky Answers

The six riddles are more than just funny catches. They teach you not to make unnecessary assumptions, but to consider all possibilities no matter how unlikely or bizarre they may seem. Some of the greatest revolutions in science would never have taken place if great minds had not questioned assumptions that everyone took for granted. Their next step—the aha! insight—was to consider a possibility that others thought crazy. For example: Copernicus guessed that the sun (not the earth) was the

习以为常的错误假设常常引导到错误的结论。

center of the solar system, Darwin guessed that mankind evolved from a lower form of animal life, and Einstein guessed that the structure of space need not conform to Euclidean geometry.

Our six sneaky riddles are answered as follows:

1. The unnecessary assumption that almost everyone makes in trying to figure out this problem is that the time is at night. But this is not stated in the problem. The room did not get dark because it was day time.

2. The false assumption is that the Aunt is of normal height. Actually, she is a midget who cannot reach high enough to push the button for her nephew's floor.

3. The false assumption is that there are other letters between the three pairs of letters. The word is "ISLAND ".

4. The false assumption is the belief that one can read only with the eyes. The man was blind, and was reading a book in braille.

5. The false assumption is that "coffee" means liquid coffee. The ring fell into a can of dry coffee, so naturally it did not get wet.

6. The false assumption is that the father had hair on his head. The father is bald, therefore he has no hair to get wet.

There are hundreds of amusing brain teasers that are based on the same basic idea—misleading one into making a false assumption that prevents one from thinking of the true explanation. Here are six more.

1. A man found a dead fly in his soup. The waiter was apologetic. He took the bowl to the kitchen and returned with what apparently was a new bowl of soup. A moment later the man called the waiter over.

"This is the *same* bowl of soup I had before!" he shouted angrily. How did he know?

2. While an ocean liner was anchored, Mrs. Smith felt too ill to leave her cabin. At noon the porthole by her bed was exactly 7 meters above the water line. The tide was raising the water line at a rate of 1 meter per hour. Assuming this rate doubles every hour, how long will it

take the water line to reach her porthole?

3. The Reverend Sol Loony announced that on a certain day, at a certain time, he would perform a great miracle. He would walk for twenty minutes on the surface of the Hudson River without sinking into the water. A big crowd gathered to witness the event. The Reverend Sol Loony did exactly what he said he would. How did he manage it?

4. Two train tracks run parallel except for a spot where they go under a tunnel. The tunnel is not wide enough to accommodate both tracks, so they become a single track for the distance of the tunnel.

One afternoon a train entered the tunnel going one direction, and another train entered the same tunnel going the opposite direction. Both trains were going at top speed, yet there was no collision. Explain.

5. An escaped convict was walking along a country road when he saw a police car speeding toward him. Before he ran into the woods he ran 10 meters directly toward the approaching car. Did he do this just to show his contempt for the police, or did he have a better reason?

6. Why are 1977 dollar bills worth more than 1976 dollar bills?

The answers are at the back of the book, but do not look at them until you have tried hard to answer each question.

The Big Holdup
大盗贼

The next day, when the discotheque's waiter reported for work, he heard shouting coming from the attic.

侦探小说往往为我们提供许多逻辑推理的难题。在这方面排除所有不可能性是其中的关键。

He rushed to the attic and found the manager with a rope around his waist and hanging from an overhead beam.
Manager: Quick. Get me down. Call the police. We've been robbed.

The manager told his story to the police.
Manager: Last night, after we closed, two robbers came and took all the money. Then they carried me to the attic and tied me to the beam.

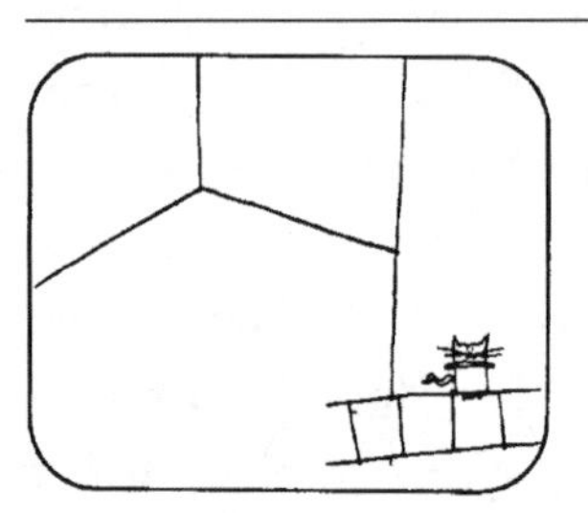

The police believed his story because the attic room was completely empty. He couldn't have tied himself to the high beam, there was nothing to stand on. There was a step ladder used by the thieves, but it was just outside the door.

However, a few weeks later, the manager was arrested for robbing himself. Can you figure out how the manager, without any help, tied himself in mid-air?

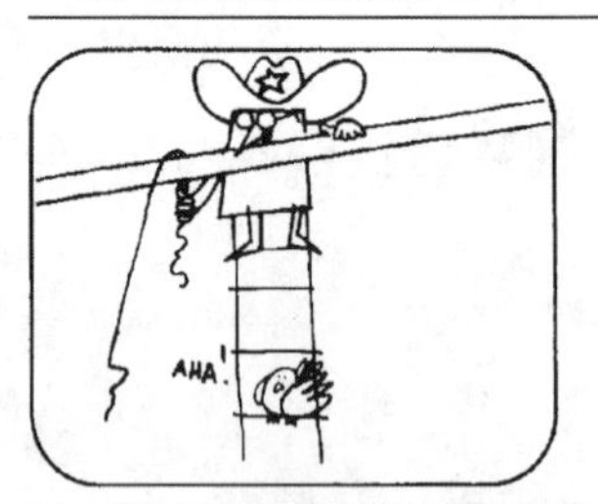

Here's how he did it. He used the ladder to tie one end of the rope to the beam. Then he carried the ladder out of the room.

He returned with a huge block of ice he had prepared in the freezer.

He stood on the ice, tied the rope around himself and waited.

When the waiter found him the next day, all the ice had melted and the manager was left hanging in mid-air. Clever, wasn't he?

Missing Evidence

Many famous mystery stories have been based on problems like this one, which the detective solves by a flash of intuition. Melting ice has been a favorite device of early mystery writers. For example, a victim is found stabbed. Where is the murder weapon? It turns out to have been a piece of ice with a sharp point like an icicle. A man is found murdered inside a room locked on the inside by a latch that had been earlier propped up with piece of ice. When the ice melted, the latch fell and locked the door.

破案离不开证据，搜集证据及寻找失掉的证据也有赖于假设与推理。此处再一次说明错误的假设往往起误导作用。如本节的录音机。

A classic puzzle mystery of this type is "The Problem of Thor Bridge" by A. Conan Doyle. A woman is found shot in the head on a bridge that has a parapet of stone on either side. There is no trace of the pistol that had fired the bullet, yet Sherlock Holmes, in a flash of insight, thinks of how the woman may have committed suicide and disposed of the weapon.

The solution is that she had tied the pistol to one end of a long cord. The string passed over the stone parapet and had a heavy stone tied to the other end. After shooting herself, the gun dropped from her hand and the stone pulled it into the water.

Holmes's solution of this problem, like so many of his others, is an excellent model of how science operates. First the great detective, by a flash of intuition, developed a *theory* to explain the disappearance of the weapon. He then deduced a consequence of the theorynamely, that the pistol striking the parapet would chip the stone. He found just such a chip mark. Finally, he devised a test to confirm the significance of the chip. He tied a stone to a string, and the other end to Watson's revolver. To simulate the suicide, he stood where the body had been found, and released the revolver. When he discovered that it made a second chip mark on the parapet, identical with the other one, his theory was amply confirmed.

This is precisely how science solves its problems. First a theory, then a deduction of practical consequences if the theory is true, then a search for the evidence and the devising of experiments to test the theory.

Here is a new mystery problem that also can be solved by a clever theory. The body of Mr. Jones was found slumped on his desk with a bullet hole through the head. Detective Shamrock Bones saw a tape recorder on Mr. Jones's desk. He pressed the play button and was surprised to hear Jones's voice saying:

"This is Jones speaking. Smith just telephoned to say that he's coming here to kill me. I'm not going to try to escape. If he carries out the threat I'll be dead in ten minutes. This recording will tell the police who killed me. I hear his footsteps now in the hallway. The door is opening ..."

There was a click to indicate that Jones had turned off the recorder.

"Shall I pick up, Smith?" asked Lieutenant Suzy Wong, who was Captain Bones's assistant.

"No," said Bones. "I'm convinced that someone else, who was good at imitating Jones's voice, killed Jones and made this recording to incriminate Smith. "

Bones's theory later proved to be correct. Can you think of what made him suspect that the recording was a fake? Try to do this before you look at the answer at the back of the book.

Dr. Ach's Tests

阿克博士的测验

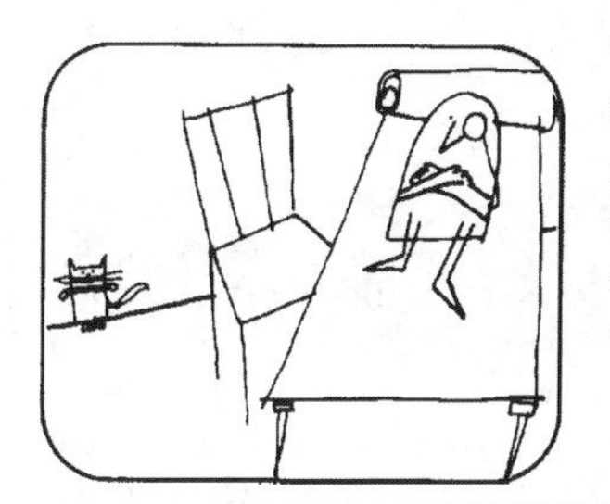

The police would never have solved the case without the help of Dr. Ach, a psychology professor who specialized in problem solving. He called his aha! insights "Ach" phenomena and devised many tests for them.

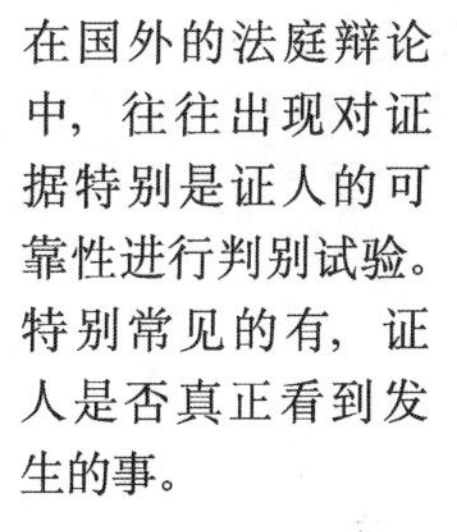

在国外的法庭辩论中，往往出现对证据特别是证人的可靠性进行判别试验。特别常见的有，证人是否真正看到发生的事。

One involved two long strings that hung from the ceiling of an empty room.

Dr. Ach: These strings are so far apart that if you hold one end, you can't reach the other.

Dr. Ach: The problem is to tie the two ends of the rope together using nothing more than a pair of scissors. Could you pass this test?

Dr. Ach: Another of my favorite tests is to put an open bottle of beer in the center of a small oriental rug. The problem is to get the beer off the rug.

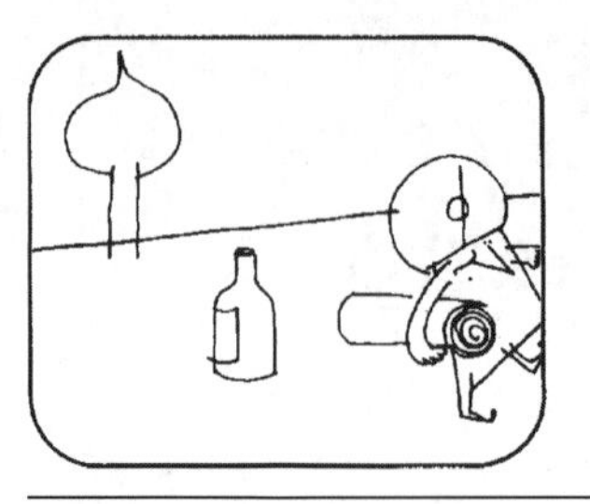

Dr. Ach: But you mustn't touch the bottle with any part of your body or anything else. And, of course, not a drop of beer must be spilled.

If you didn't pass the last test, maybe you'll get this one.

Dr. Ach: For my last test you need a sheet of newspaper. The problem is for you and a friend to stand on it in such a way that you can't touch each other.

Naturally you can't step off the paper. This is your last chance to pass one of Dr. Ach's tests.

Doctor Ach doesn't like to remember that test because one of his students answered it and challenged him with another.

Student: Alright, Dr. Ach, try to throw this tennis ball so it goes a short distance, comes to a dead stop, then reverses itself and goes the opposite way.

Dr. Ach: May I bounce it against something?

Student: No, bouncing is not allowed. And you can't hit it with anything or tie anything to it either.

After Dr. Ach gave up, the girl surprised him by taking the ball and doing exactly what she said.

Dr. Ach: Why didn't I think of that?

What was it that he failed to think of?

Dr. Ach's Solutions

解决问题的关键都是破除“定势思维”方式。

Dr. Ach's Strings: You may think one could solve this by grabbing a string and swinging on it like Tarzan in the manner shown in one of the pictures. This would not work for two reasons: The string is not strong enough to support a person, and even if it were, the person could not reach the other string. However, the picture does give a clue to the correct solution.

If you tie the scissors to one end of a string, you can set the string swinging like a pendulum. This allows you to pull the end of the other string as near as possible to the swinging string, and catch the scissors when they swing toward you.

It takes two insights to solve this task problem. One is to think of swinging strings, and the second is to think of using the scissors in a way for which they were not designed. Psychologists have a term called "functional fixedness" for the difficulty people have in using devices in unaccustomed ways. The mind thinks only of how scissors can *cut* string. Of course, cutting a string is no help in solving the problem.

Dr. Ach's Rug: You are not allowed to touch the bottle with any part of your body, or with anything else. The insight that solves this task is the realization that, since the rug is *already* touching the bottle, perhaps the rug itself can be used for moving the bottle off the rug.

This proves to be true. Merely roll up the rug at one end. When you get to the bottle, roll it slowly with your hands at each end, and the middle of the roll will slowly push the bottle off the rug without tipping the bottle over.

As in the previous problem, functional fixedness is a mental block to the solution. One thinks of a rug only as a floor covering, not as an object that can be used as a pushing tool.

Dr. Ach's Newspaper: The aha! that solves this task is the realization that a door separates two persons who stand on the same sheet of newspaper. Simply put the sheet under an open door. The boy stands on the paper at one side of the door, and the girl stands on the other side. The door prevents them from touching each other without stepping off the paper.

Tennis Ball: The mental block here is the assumption that the ball is tossed horizontally. But there is nothing in the statement of the problem to prevent one from tossing the ball straight up in the air. Naturally it comes to a complete stop, reverses its motion, and goes the opposite way!

Another solution is to roll the ball up a hill. This could have been ruled out by stating that the ball must travel through the air without touching anything, but since we didn't say this, it counts as a legitimate solution.

More Task Problems: Here are six more task problems that you and your friends will enjoy. Try to solve them before reading the answers.

1. Can you drop a paper match from a height of about 1 meter so that it falls on its edge and remains on its edge?

2. Some workmen are making mortar with sand and cement so they can lay the foundations of a building. One of the large concrete blocks has a small rectangular hole 2 meters deep. A baby bird has fallen into the hole. The hole is too narrow for an arm to be squeezed into it; besides, the bird is too deep to be reached by an arm. Trying to grasp the bird with two sticks would injure it. Can you think of a simple way to get the bird out of the hole?

3. Tie one end of a piece of string, about 2 meters long, to the handle of a coffee cup. Tie the other end to a hook in the ceiling, or over an open doorway, so that the cup hangs suspended. The problem is to cut the center of the cord with a pair of scissors so that the cup will not fall to the floor. No one may hold the cup, or touch the string while it is being cut.

4. A dike in Holland is missing a single brick. Water is pouring through the rectangular hole which is 5 centimeters by 20 centimeters. The man who discovered the hole has with him a saw and a cylindrical wooden pole with a diameter of 50 millimeters. What is the best way he can cut the pole so as to plug up the hole?

5. A wine bottle has a cylindrical shape for its lower section. The section is 3/4 of the bottle's height. The upper fourth of the bottle is irregularly shaped. The bottle is filled half-way up. Without opening the bottle, and with the aid of only a ruler, how can you determine exactly what percentage of the bottle's total volume is filled?

The answer to these five problems appear at the end of the book.

The Ach Award

阿 克 奖

At the end of every course in Ach thinking, Dr. Ach gave a special Ach medal to his best student. One year there were three students tied for the honor.

本节问题是更好的逻辑问题，它比本章第2节颜色配对更进一步。

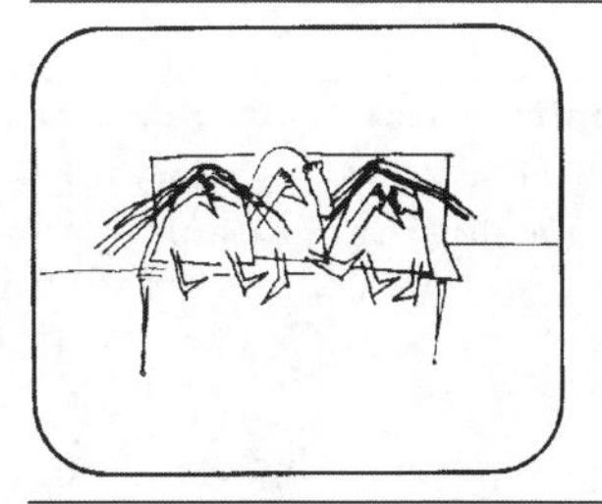

Dr. Ach used a test to break the tie. He seated the three students on a bench and told them to close their eyes.

Dr. Ach: I'm going to put a red or a blue hat on each of you. But don't open your eyes until I tell you.

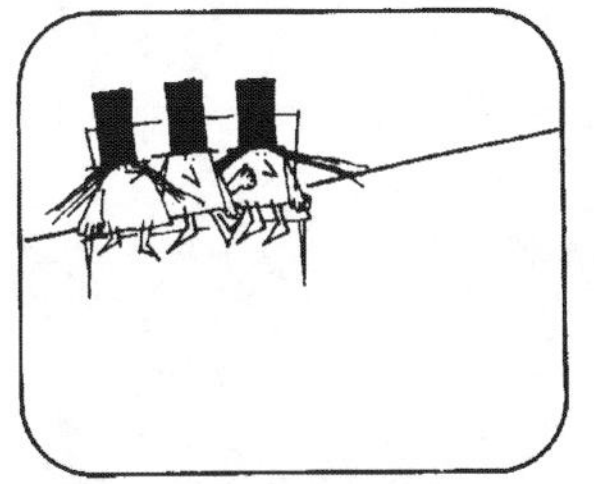

Dr. Ach put a red hat on each of them.

Dr. Ach: Now open your eyes and raise your hand if you see a red hat on someone. The first person to deduce the color of his own hat gets the medal.

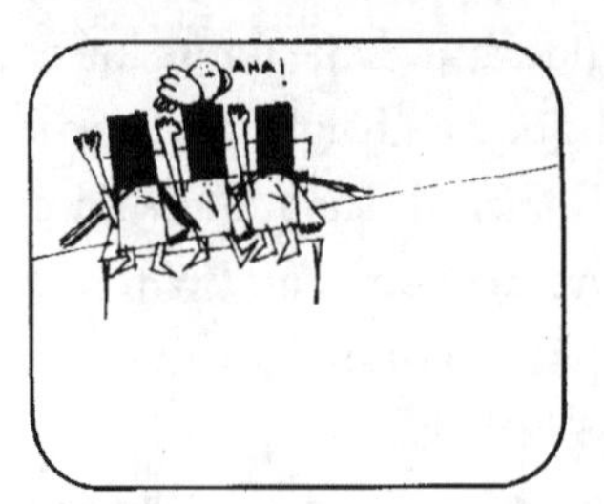

Of course, all three raised their hands. But several minutes passed before John stood up and shouted.

John: Ach, I know my hat is red.

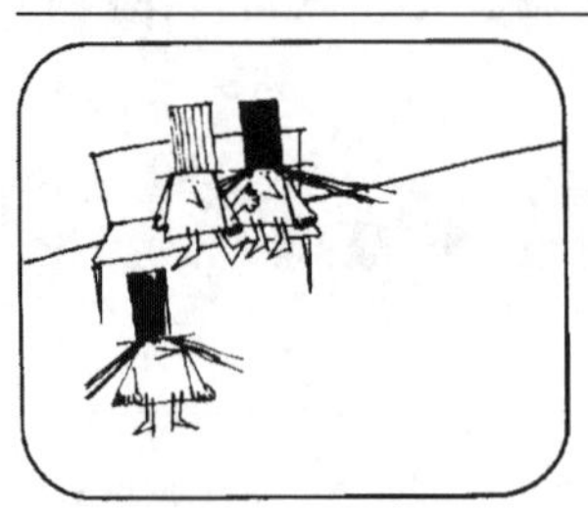

John: If my hat were blue, Mary would know at once that her hat was red because that would be the only way to explain Barbara's raised hand.

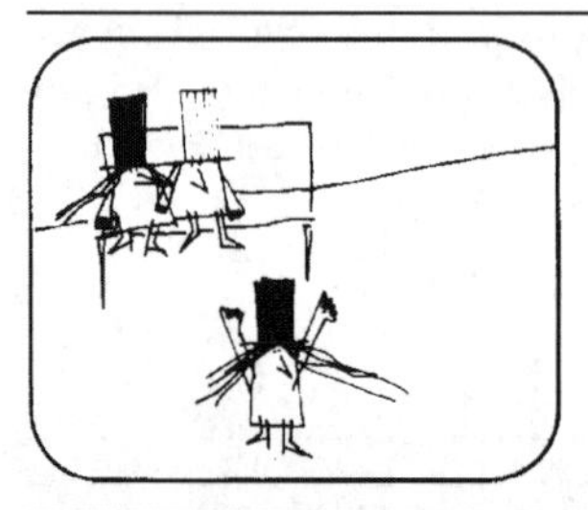

John: Naturally Barbara would think the same way. She would know that her hat was red because that would be the only way to explain Mary's raised hand.

John: But neither girl could name the color of their own hat. So they must be seeing a red hat on me too.

It's easy to understand this classic logic puzzle when there are only three persons. But suppose there are four, and all get red hats. Can you figure out what would happen?

▥ blue
■ red
▦ green

Inductive Color Thought

Going from three persons in this problem to four, then generalizing to any number of persons, is an excellent introduction to a valuable technique of proof called "mathematical induction" . It applies only when there are statements that can be ordered like the rungs of a ladder. You first show that any statement is true if the previous one is true. If the first is true, then all the others must be true. If you can step on the first rung of the ladder, you can climb it all the way to the top. Or if you start on a higher rung, you can climb all the way up or down.

这个问题特别有趣，即除了问题本身之外，还有一个元问题：每人都不确定自己头上的帽子的颜色。只有知道大家都不确定的情况下，某个聪明人才能正确推断出自己头上帽子的颜色。

Suppose there are four men who all get red hats. All raise their hands. Assume that one of them has more aha! insight than the others. He or she reasons as follows:

"Suppose my hat is blue. The other three will all see that it is blue. Each person, therefore, will see two red hats and wonder about his own. But this is exactly the situation of the previous problem when there are just three people. Eventually one of the three will deduce that his hat is red.

"However, suppose enough time has elapsed for such a deduction, but no one has made it. There can be only one reason, and that is because they all see that my hat is also red. Therefore my original assumption is false. My hat must be red. "

This generalizes to n persons. If there are five persons with red hats, the cleverest will see four red hats and realize that after a sufficient time one of the four persons has to reason as explained above, and know that his/her hat is red. But if no one is able to do this, it indicates that his/her own hat must also be red. And so on for any number of persons. The cleverest among n people can always reduce the situation to the previous case, which in turn reduces to the previous case, and so on back down to the case of three persons, which is solved.

The general problem can lead to interesting arguments about whether it is sharply defined, or whether it is too ambiguous in its conditions to have a sharp answer. What assumptions must be made to make the general solution valid? Is it necessary that the reasoning abilities of the *n* persons form a hierarchy? Is it necessary to assume that as *n* increases, the length of time it takes for a person to deduce his/her hat is red also increases? Is it correct to say that if there are 100 persons, then after a very long length of time the cleverest will know that his/her hat is red, then after another lapse of time, the second cleverest will know, and so on down to the least clever last man or woman?

There are endless variants of the classic hat problem. Here is one that shows how the problem can be complicated by introducing hats of more than two colors. Suppose that five men are given hats selected from a set of five white, two red, and two black. If all the hats are white, how does one man, cleverer than the rest, deduce that his hat is white?

A particularly elegant three-person variant of the original two-color problem eliminates all of its ambiguities. Assume that three men are seated in three chairs, one behind the other and all facing the same way. The man in the back chair can see the hats of the two men in front. The man in the middle sees only the hat of the man in front. And the man in front sees no hat. Think of the men as progressively "blind", with the man in the front chair totally blind.

An umpire picks three hats from a set of three white and two black. The men close their eyes until the hats are placed, and the unused hats concealed.

The umpire asks the man in back if he knows the color of his hat. He replies "No."

The man in the middle is asked the same question. He, too, says "No."

When the man in front is asked, he replies, "Yes, my hat is white." How did he deduce this?

He reasoned as follows: "The man in the back chair will say 'yes' only if he sees two black hats. His 'no' answer proves that the two hats he sees are not both black. Suppose now that my hat is black. The man in the middle sees it is black. As soon as the middle man hears the man behind him say 'no', he knows that his own hat must be white—otherwise he would see two black hats and say 'yes'. Therefore, the middle man would say 'yes.' However, he actually said 'no.' This proves that the middle man sees a white hat on my head. Therefore, my original assumption is false, and my hat is white. "

Like the earlier version, this also generalizes easily by mathematical induction to n "progressively blind" men seated in a row of n chairs. Questions start with the man in back, then go forward. The supply of hats consists of n white hats and n-1 black. Consider the case of $n = 4$. The "blind" man in front knows that if his hat is black, the three men behind him see his black hat and know that only two black hats are left for themselves. This reduces the problem to the previous case. If the first two men say 'no,' the third man (seated directly behind the blind man) would then say 'yes', as in the previous case. If, however, he says 'no', it proves to the blind man that his assumption is false, and his hat must be white. Mathematical induction then extends the proof to n persons. If all but the blind man say 'no', all must be wearing white hats.

A more difficult question can now be asked. Suppose that in the three-man case the umpire gives them *any combination* of hats from the set of five (three white, two black). The men are questioned in the same order as before. Will one of them always answer yes? You may enjoy working this out, and proving that it generalizes to n persons and a set of n white and $n - 1$ black hats. Someone will always answer yes. The first person who does is always the first one asked who is wearing a white hat and who sees no white hat in front of him.

Hats of two colors are equivalent to hats labeled 0 and 1, the integers of binary notation. There are many hat problems involving more than two colors (such as the one given earlier), but they are easier to understand if instead of colors we use positive integers. Consider, for example, the following two-person game.

The umpire chooses any pair of consecutive positive integers. A disk with one of the numbers is stuck on the forehead of one player, and a disk with the other number is stuck to the forehead of the other player. Each sees the other's number, but not his own. Both men are honest and rational.

数学归纳法的重要性。许多问题可以推广到一般情形。

The umpire asks each man if he knows his number, and the questioning continues back and forth until one man says yes. Using mathematical induction, you can show that if the higher of the two numbers is n, one man will say yes to question n or question $(n-1)$. The proof starts by considering the simplest case: Numbers 1 and 2. The man with 2 will say yes to the first or second question (depending on who is asked first) because, seeing 1, he knows his number is 2.

Now consider the 2 and 3 cases. The first time the man with 3 is asked, he will say no because he could be 1 or 3. Assume he is 1. In that case the man with 2 would say yes (as in the previous case). Consequently, if he says no, this proves to the other player that his number is 3, rather than 1, therefore he says yes when asked the second time. As in the hat problems, this generalizes for any pair of consecutive numbers.

For the full solution you need to know just when a player will say yes on question n, and when on $(n-1)$. You will find that this depends on which man is asked first, and whether n is odd or even.

A more sophisticated generalization has been investigated recently by the famous Cambridge mathematician John Horton Conway. It goes like this. Numbered disks are placed on the foreheads of n men. The numbers can be any set whatever of non-negative integers. The sum of all these integers is one of n or fewer numbers written on a blackboard. The blackboard numbers must have no two alike. The men are assumed to be infinitely intelligent and honest. Each can see all the disks except his own, as well as all the blackboard numbers.

The first man is asked if he can deduce the number on his forehead. If he says no, the question is asked of the second man, and this questioning continues cyclically around the men until one of them says yes. Conway asserts that, incredible as it may seem, the questioning always terminates with a yes.

Holiday Haircut

假日理发

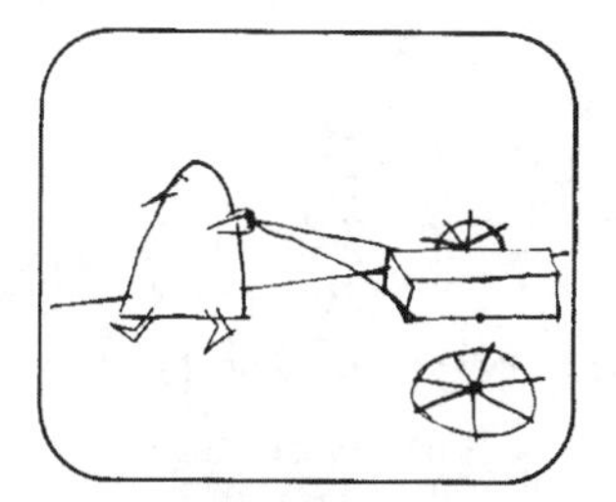

John was driving to Las Vegas for a vacation when his car broke down in a small town. While the car was being fixed John decided to get a haircut.

这问题涉及正确的选择标准以及标准的自适用性。

The town had just two barber shops, Joe's and Bill's.

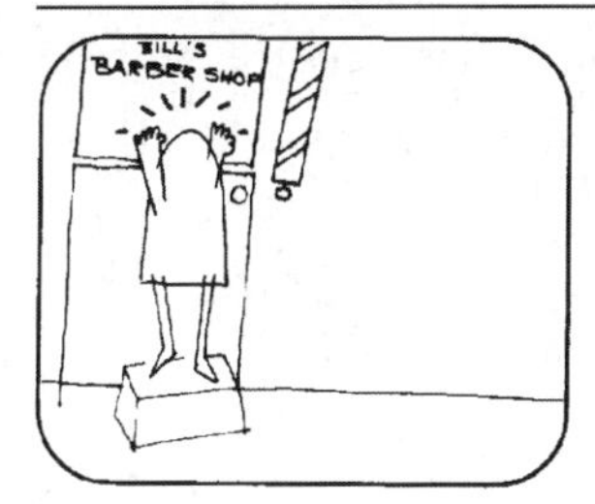

John looked through the window of Bill's shop and was disgusted.
John: What a dirty shop. The mirror needs cleaning, there's hair all over the floor, the barber needs a shave, and he has a terrible haircut.

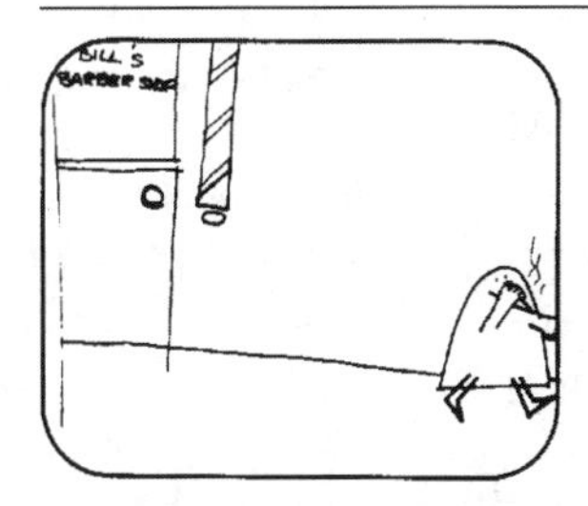

It's no wonder that John left Bill's shop and went up the street to check on Joe's barber shop.

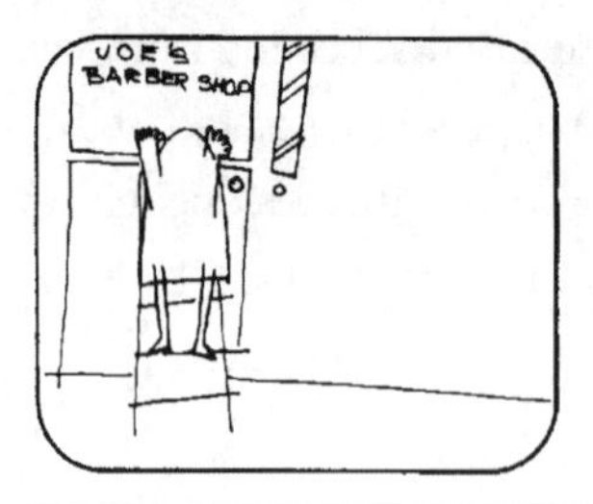

John peeked through Joe's window.
John: What a difference. The mirror is clean, the floor's clean, and Joe's hair is neatly trimmed.

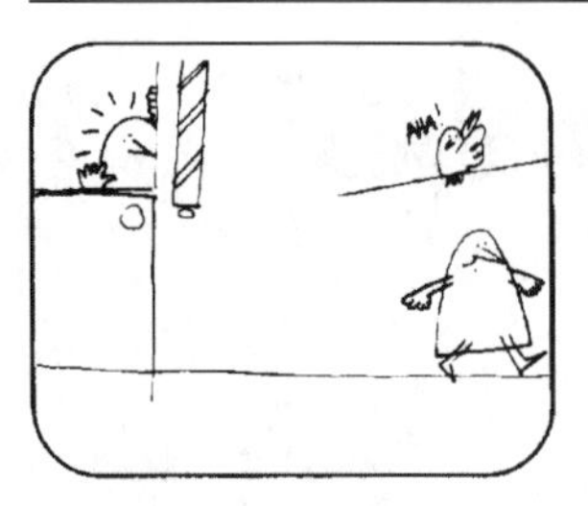

But John didn't go in. Instead, he walked back to get his hair cut at Bill's dirty shop. Why?

Which Barber?

罗素悖论的通俗形式就是理发师悖论。它就涉及一个命题是否可适用于自身。

No barber cuts his own hair. Since the town has only two barbers, each must have to cut the hair of the other. John wisely had his hair cut in the dirty barbershop because its barber had given such a neat haircut to the owner of the clean shop.

A problem very similar to this one goes as follows. Two miners, who had been working all day in a coal mine, finished their work and came up to the surface. One of them had a clean face, the other had a face covered with coal dust. They said goodnight to one another. The man with the clean face wiped his face with his handkerchief before he started home, but the man with the dirty face did nothing to his face. Can you explain this peculiar behavior?

Barbershop Bantor

理发店的玩笑

问题必须确切，不能有两种解释，一旦作出正确的严格解释，又不违背题意，问题即可解决。

Bill was a talkative barber and could hardly wait to get started.
Bill: So you're from out of town, hm hmm? I like to cut the hair of strangers.

Bill: In fact, I'd rather cut the hair of two people from out of town than the hair of anybody who lives here.
John: Why is that?

Bill: Because I get twice as much money.

John: Okay, you caught me on that one. But here's one for you. Ten days ago our college basketball team won a game with a score of 76 to 40. And yet not one man on our team got a basket. Can you tell me why?

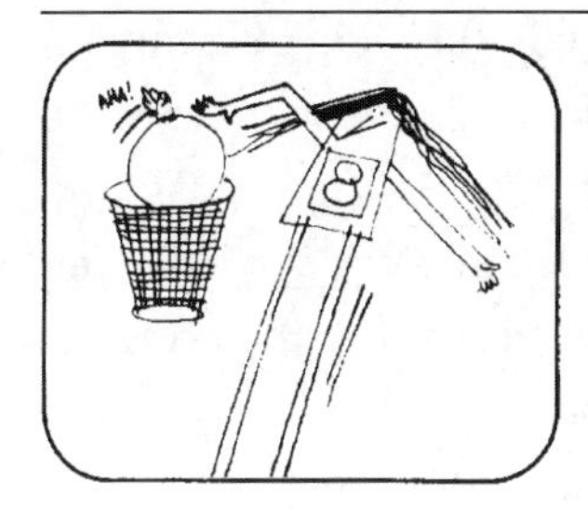

The barber was stumped, so John explained.
John: There aren't any men on our team. They're all girls.

Surprising Solutions

The problems in this section are all humorous "catches" based on verbal ambiguity. Here are eight more problems of the same type to catch your friends with.

1. Howard Youse, an eccentric billionaire, offered a prize of half a million dollars to the racing car driver whose car came in *last* in a race. Ten drivers entered the contest, but were puzzled by Mr. Youse's conditions.

"How can we run the race?" one of them asked. "We'll all just go slower and slower, and the race will never finish."

Suddenly one of them said, "Aha! I know how we can manage it." What did he think of?

2. How can you make a match burn under water?

3. A criminal took his wife to a movie theater that was showing a shoot-em-up western. During one scene, when many guns were fired, he murdered his wife by shooting her in the head. He then took his wife's body out of the theater, but no one stopped him. How did he manage it?

4. Professor Quibble says he can put a bottle in the center of a room and crawl into it. How does he do it?

5. Uriah Fuller, the famous Israeli superpsychic, can tell you the score of any baseball game before the game even starts. What is his secret?

6. A man who lived in a small town married twenty different women of the same town. All are still living, and he never divorced a single one of them. Yet he broke no law. Can you explain?

7. "This myna bird," said the pet shop salesman, "will repeat any word it hears." A week later the lady who bought the bird was back in the shop to complain that the bird had not yet spoken a single word. Yet the salesman told the truth. Explain.

8. A wine bottle is half filled and corked. How can you drink all the wine without breaking the bottle or removing the cork from the bottle?

The answers are at the back of the book.

Murder at Sun Valley

太阳谷的谋杀者

这个问题涉及侦探破案另一方面，即动机问题。作案动机可以有许多假定，必须有如题中的证据才能确认，否则可能会引发误导。

When John got to Las Vegas the papers were headlining a story about a local gambler and his wife who had been skiing at Sun Valley.

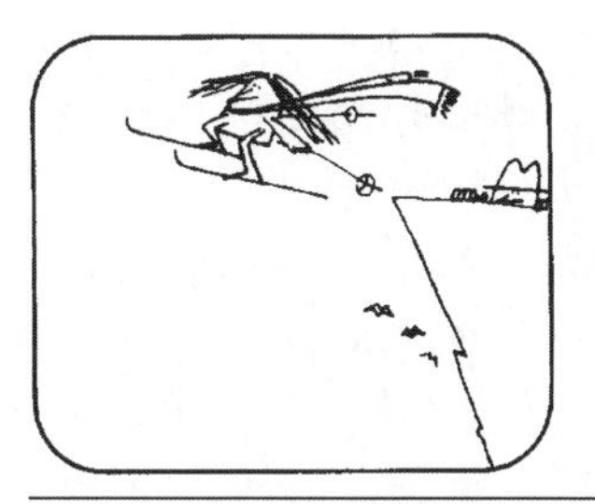

The wife had died after a skiing accident. The gambler had been the only witness to her fall when she skidded over a precipice.

A clerk in Vegas read about the accident and phoned the Idaho police. The gambler was arrested on suspicion of murder.

The reporters were surprised with the clerk's story.

Clerk: I don't know the gambler or his wife and I didn't suspect foul play until I read about the accident.

Why then, did the clerk call the police?

Because he had sold the gambler a round trip ticket to Sun Valley, but only a one way ticket for his wife.

The One Way Ticket

Now see how well you do on these two mystery problems. Like the previous one, they cannot be solved by any kind of algorithm or planned procedure, but a correct aha! reaction leads quickly to the answers.

思维定势往往使我们不能涵盖所有的情形。例如，外科医生可以是女性。

1. On a thruway to San Francisco, a father was driving with his small son in the front seat. He swerved to avoid hitting a stalled car, lost control of his car, and smashed into a bridge abutment. The father was unhurt but the boy suffered a broken leg.

An ambulance took them to a nearby hospital. The

boy was wheeled into the emergency operating room. The surgeon was about to operate. Suddenly the surgeon cried out: "I can't operate on this boy. He's my son! " Explain.

2. The following story is adapted from *Puzzle-Math*, a delightful collection of problems by George Gamow and Marvin Stern. At the time of the German occupation of France, during World War II, four people were riding a hotel elevator in Paris. One passenger was a Nazi officer in uniform, another was a native Frenchman who was a secret member of the underground. The third passenger was a young, pretty girl, and the fourth was an elderly lady. They were all strangers to one another.

Suddenly a power failure occurred. The elevator stopped and the lights went out, leaving the car in total darkness. There was the sound of a kiss, followed by the sound of a punch in the face. A moment later the power was back on. The Nazi officer had a fresh bruise under one eye.

The elderly lady thought: "Serves him right! I'm glad that young girls these days know how to take care of themselves. "

The young girl thought: "What strange tastes these Nazis have! Instead of kissing *me*, he must have tried to kiss this older woman or this nice young man. I can't figure it out! "

The Nazi officer thought: "What happened? I didn't do *anything*. Maybe this Frenchman tried to kiss the girl and she hit me by mistake. "

Only the Frenchman knew exactly what had happened. Can *you* deduce what took place?

Both problems are answered at the back of the book; but try to solve the problems first! (before looking).

Foul Play at the Fountain

喷泉边的谋杀

此题与逻辑推理关系不太密切。

John checked into a Vegas hotel on the strip and while he was reading a newspaper, a gorgeous girl rushed into the lobby.

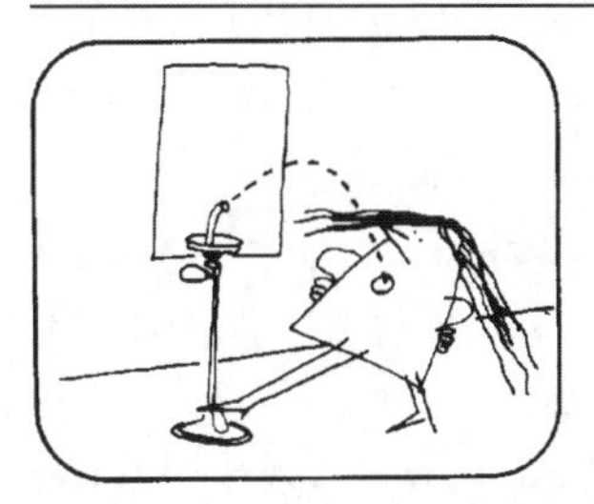

Then she ran over to the water fountain, took a long drink, and disappeared.

Three minutes later the same girl came back for another long drink. This time she was followed by a strange looking man.

There was a mirror in back of the fountain. And when she raised her head she saw the man standing behind her with a big knife in his upraised fist as if to stab her in the back.
Lady screams.

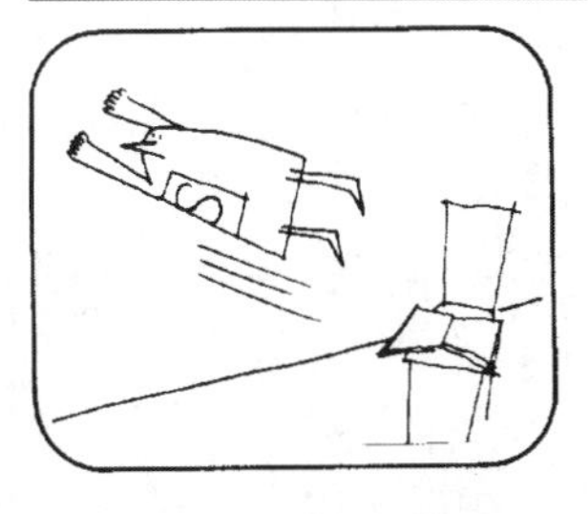

John leaped to the rescue.

But then the man lowered his knife and he and the lady began to laugh. What on earth is going on?

Mirror Vision

多少带智力测验性质。

The strange behavior of the lady is easily explained. She had the hiccups, and the man was trying to frighten her out of them.

Now for a last chance to test your aha! logic ability. First an operational task problem, then a clever question based on an unwarranted assumption.

1. Cleopatra keeps her diamonds in a box with a sliding lid on top. To foil thieves she has a live and deadly asp inside the box with the jewels.

One day a slave was left alone in a room with the box for only a few minutes. He managed to steal several priceless gems without taking the snake out of the box and without touching or influencing the snake in any way. He wore nothing to protect his hands. The theft took only a few seconds. When the slave left the room, the box and snake were in exactly the same condition as before except for the absence of several diamonds. What ingenious method did the slave use to steal the gems?

2. A lady did not have her driver's license with her. She failed to stop at a railroad crossing, then ignored a one-way traffic sign and travelled three blocks in a wrong direction down the one-way street. All this was observed by a policeman, yet he made no effort to arrest the lady. Why?

The answers are at the back of the book.

Chapter 5

Procedural aha!

过　　程

Puzzles about routines

Since the computer revolution began, the word "algorithm" has become a familiar term in the lexicon of mathematics. It simply means a procedure—one made up of a series of well-defined steps—that will solve a problem. When you divide one big number by another, you do it by using a division algorithm. Since computers cannot solve a problem without being told exactly what to do, the art of computer programming is mostly the art of constructing efficient algorithms. We say "art" rather than "technique" because the mysterious aha! plays a significant creative role in the discovery of good algorithms.

解决任何问题都涉及一系列步骤，对数学来说，这一系列确定的步骤称为算法。计算机的程序设计就是构造有效的算法。

当一个问题还没有算法时，首要问题是设计一个算法。当已经有算法时，问题就是如何简化这个算法。在这个过程中，把一个问题化为已有算法的问题是十分重要的。

By "good" we mean an algorithm that solves a problem in the shortest time. It costs money to run a computer, just as it costs money to hire laborers for a job. As a result, there are great practical advantages in having efficient (good) algorithms. Indeed, a flourishing branch of mathematics called O. R. (Operations Research) is concerned explicitly with finding the most efficient ways for solving complicated problems.

Although the procedural problems in this section have been selected because they are entertaining, you will find yourself painlessly learning many deep mathematical concepts. The first puzzle, for example, brings out vividly what mathematicians mean when they say two seemingly unrelated problems are "isomorphic". A carnival betting game involving numbers turns out to have a strategy with a structure that is identical with the strategy of playing tick-tack-toe! This, in turn, is shown to be isomorphic with a clever word game invented by the Canadian mathematician Leo Moser, as well as a game played on a network. And all these games have strategies based on the 3-by-3 magic square, one of the most ancient of all combinatorial curiosities.

Other tie-ins with significant concepts include: Archimedes' law of floating bodies that solves a hippopotamus-weighing task; an unsolved general problem in decision theory that follows from a simple task of dividing household chores ; some classic combinatorial problems suggested by the thief and bell rope; and important graph theory problems suggested by the problem of "The Lazy Lover ".

Graph theory is the study of sets of points that are joined by lines. Many practical problems in operations research can be modeled by graphs. Some have elegant solutions, such as the minimal spanning tree that we learn how to construct by "Kruskal's algorithm". We consider another closely-related problem, known as "Steiner's tree problem", that is still unsolved in general. Since Steiner trees have so many practical applications, a great deal of work is now going on in the search for efficient computer algorithms that find such trees.

Steiner's problem belongs to a fascinating class of problems known as NP-complete. These are problems that are unsolved in the sense that no good algorithms for them are known, nor is it known if such algorithms exist. The best known algorithm for finding a Steiner tree for n points is such that, as n increases, the time required for finding the tree grows exponentially. Indeed, it grows so fast that even for a relatively small number of points (say a few hundred) a computer might take millions of years to produce the best answer!

NP-complete problems are related to one another in a curious way. If an efficient computer algorithm is found for one of them it can immediately be applied to all the others. And if any one of the algorithms is shown to be such that there *is* no efficient algorithm, this also settles the matter at once for all the others. Mathematicians suspect that the latter is true. A great amount of work is now going on in searching for efficient algorithms that will find, not the best Steiner tree, but one reasonably close to the best.

This section, more than any other in the book, opens vistas on current research that is being done by some of the top minds in modern mathematics.

Fifteen Finesse

十五的技巧

许多数学游戏就像下棋和打牌。在一定的游戏规则之下，谋求先胜。许多游戏总结在当代著名数学家John Horton Conway等人写的*Winning Ways*（《取胜之道》）中，本节选的问题就是一些最简单的。

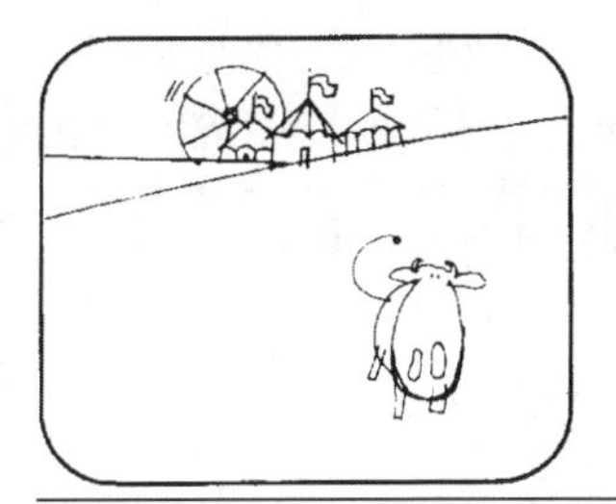

When a country fair opens, everybody gets excited. That is everybody except the cows.

This year, there's new game called "Fifteen" on the carnival midway.

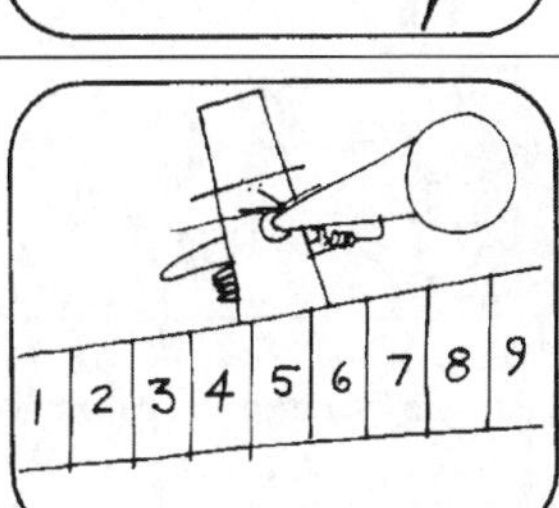

Mr. Carny: Step right up folks. The rules are simple. We just take turns putting down coins on these numbers from 1 to 9. It doesn't matter who goes first.

Mr. Carny: You put down nickels, I put down silver dollars. Whoever is first to cover three different numbers that add to 15 gets all the money on the table.

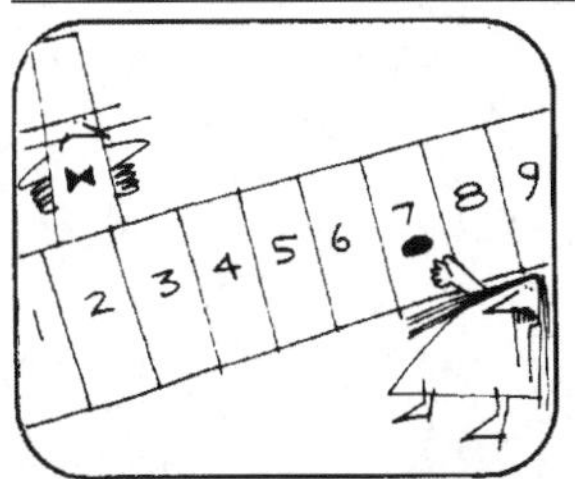

Let's watch a typical game. This lady goes first by putting a nickel on 7. Because 7 is covered, it can't be covered again by either player. And it's the same for the other numbers.

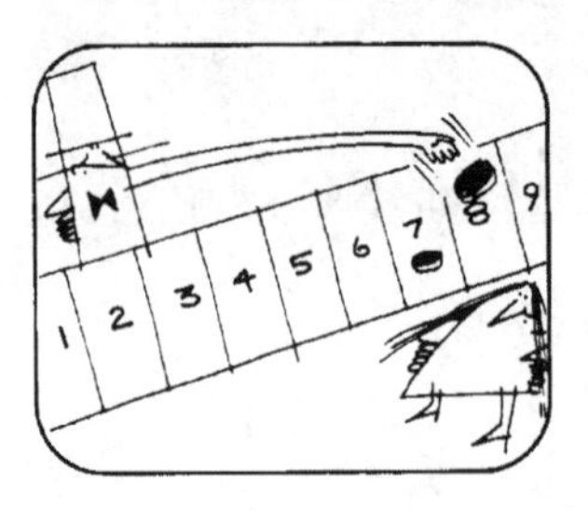

The carnival man puts a dollar on 8.

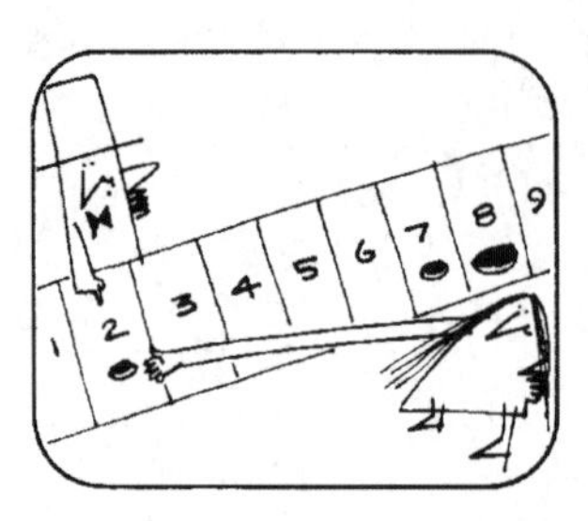

The lady's next move is to put a nickel on 2 so that one more nickel on 6 will make 15 and win the game for her.

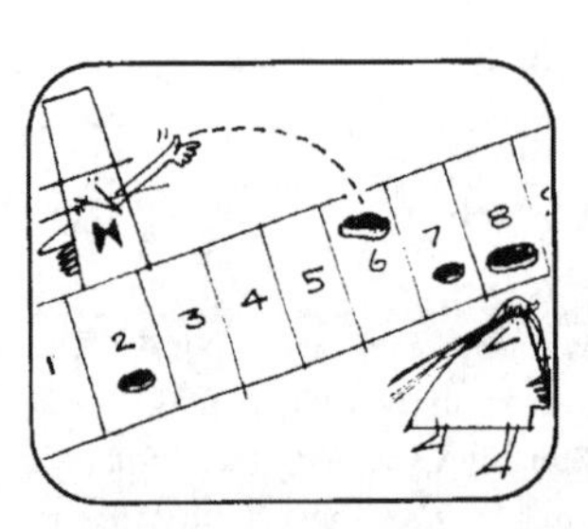

But the man blocks her with a dollar on 6. Now he can win by playing a 1 on his next turn.

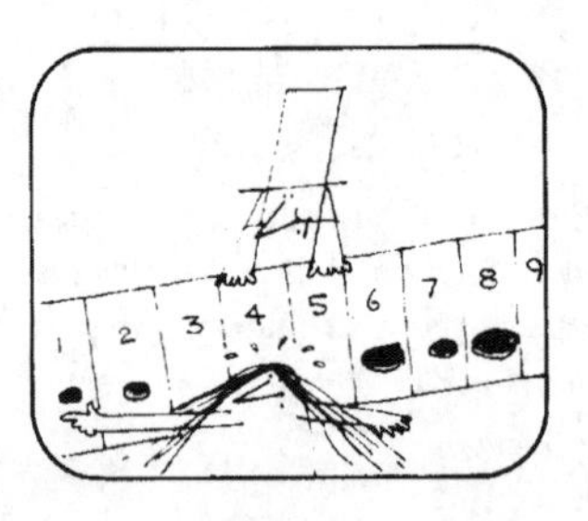

The lady sees the threat and blocks his win with a nickel on 1.

The carnival operator is chuckling as he places his next dollar on 4. The lady, seeing that he can win by playing on 5 next, has to block him again.

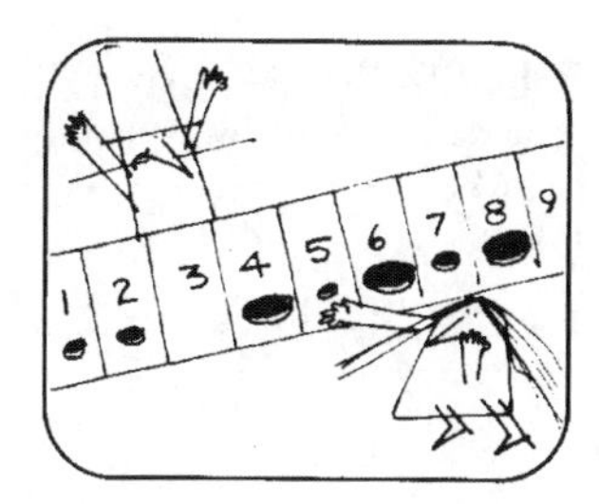

So she puts a nickel on 5.

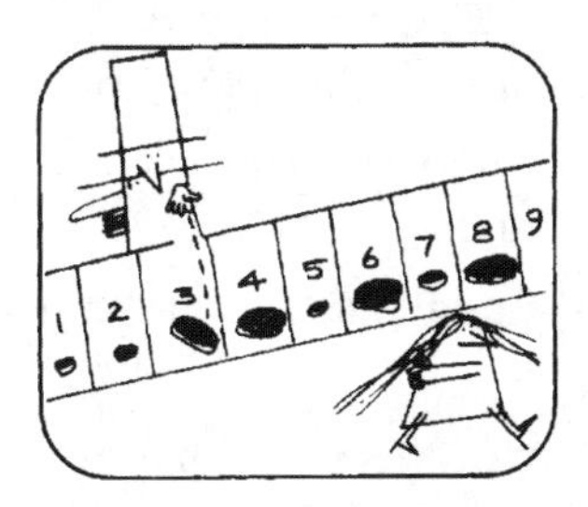

But the man now places a dollar on 3 and wins because 8 and 4 and 3 add to 15. The poor lady has lost four nickels.

His Honor, the town's Mayor, was fascinated by the game. After watching it for a long time he decided that the carnival man was using a secret system that made it impossible for him to lose except when he wanted to.

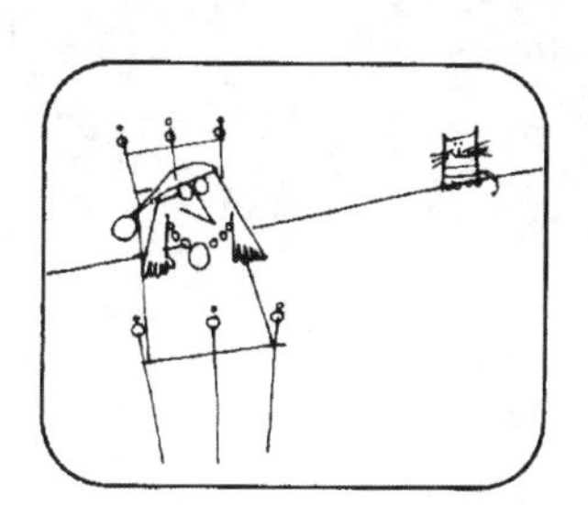

The Mayor was awake all night trying to figure out the secret system.

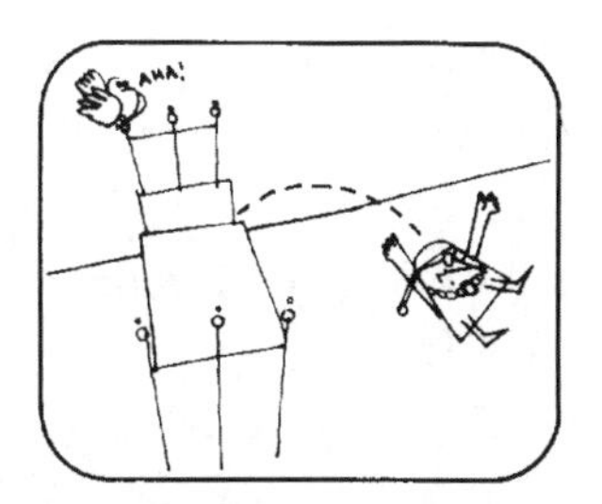

Suddenly he leaped out of bed.
Mayor: Aha! I knew he had a system. I know how he does it now. It really is impossible for a customer to win.

What insight did the Mayor have? Maybe you can discover how to play fifteen with your friends and never lose a game.

tick-tack-toe 是在一个 # 号的 9 个格子中，双方轮流画上自己的符号，谁的符号先连成一条直线谁算获胜。具体玩法见图 1。这个井字游戏和上面的 15 游戏都很简单，但最重要的是它们与幻方的关系。

幻方是起源于中国的最古老的教学游戏，也是最好的数学游戏。之所以说它好，就是玩幻方不需要有多少数学知识，马上即可入手做。而且只要有兴趣永远有做不完的事，而且有创造发明的余地。

不同的游戏一个套路称为“同构”。这是本书给出的最重要概念之一。不仅井字游戏和 15 游戏与 3 × 3 幻方有关，而且图 4 的画道路游戏玩法也是一样。

Tick-tack-toe!

The insight that solves the 15 game is the recognition that it is mathematically equivalent to tick-tack-toe! Surprisingly, the equivalence is established by way of *lo-shu*, the well known 3-by-3 magic square that was first discovered in ancient China.

To appreciate the elegance of this magic square, first list all combinations of three digits (no two alike and excluding zero) that add to 15. There are just eight such triplets :

$$1+5+9=15$$
$$1+6+8=15$$
$$2+4+9=15$$
$$2+5+8=15$$
$$2+6+7=15$$
$$3+4+8=15$$
$$3+5+7=15$$
$$4+5+6=15$$

Now look closely at the unique 3-by-3 magic square:

2	9	4
7	5	3
6	1	8

Note that there are eight sets of cells that each lie on a straight line: The three rows, the three columns, and the two main diagonals. Each of these straight lines identifies one of the eight triplets that add up to 15. Therefore, each winning set of three digits in the carnival game is represented on the magic square by a row, a column, or a diagonal.

It is now easy to see that every carnival game is equivalent to a game of tick-tack-toe played on the magic square. The carnival operator has the *lo-shu* drawn on a card that he can see (but no one else can) by looking below the playing table. There is only one *lo-shu* pattern, but, of course, it can be rotated to four different positions, each of which can be mirror-reflected to make four other forms. Any one of these eight forms is as good as any other to use as the secret key for playing the game.

As the 15 game proceeds, the carnival operator mentally plays a corresponding game of tick-tack-toe on his secret card. If one plays tick-tack-toe correctly, it is impossible to lose. If both players play correctly, the game is a draw. However, players of the carnival game are at an enormous disadvantage because they do not realize they are playing tick-tack-toe. This makes it easy for the operator to set up traps that are winning positions.

To see exactly how this works, let's play through the game shown in the pictures of this section. The moves are shown in Figure 1. Even though the carnival man went second, he was able to set a trap on move 6 that gives him a sure win on move 8 regardless of how the lady plays on move 7. Anyone who learns to play a perfect game of tick-tack-toe can, with the aid of the magic square, play an unbeatable game of 15.

1

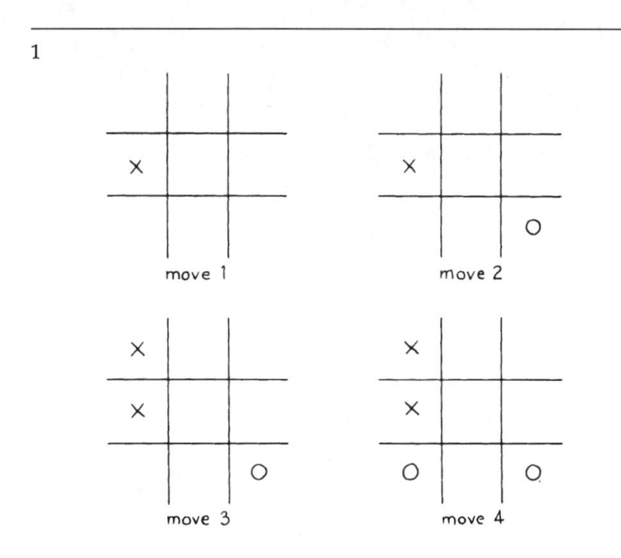

1 (continued)

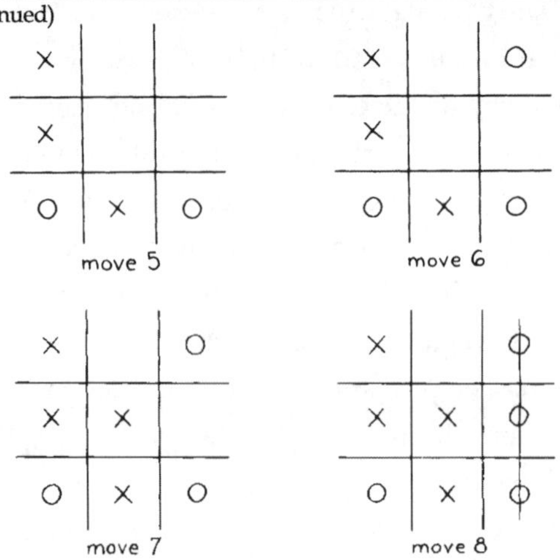

X		
X		
O	X	O

move 5

X		O
X		
O	X	O

move 6

X		O
X	X	
O	X	O

move 7

X		O
X	X	O
O	X	O

move 8

The concept of isomorphism (mathematical equivalence) is one of the most important ideas in mathematics. There are many cases in which a difficult problem can be solved easily by transforming the problem to an isomorphic one that has already been solved. As mathematics grows more complicated, it also at the same time grows more unified in the sense that it is simplified by the discovery of isomorphisms. For example, when the famous four-color map theorem was proved true in 1976, it simultaneously proved true dozens of other important conjectures, in other branches of mathematics, that were known to be isomorphic with the four-color theorem.

To develop further your understanding of the fundamental concept of isomorphism, consider the following word game. It is played with these nine words:

HOT
TANK
TIED
FORM
HEAR
BRIM
WOES
WASP
SHIP

Two players take turns crossing out a word and initialing it. The first player who crosses out three words that have the same letter in common is the winner. It may take a lot of playing before one realizes that he is simply playing tick-tack-toe! The isomorphism is easy to see by writing the words in the cells of a tick-tack-toe board as shown in Figure 2. A careful inspection shows that every triplet of words with one letter in common is in a straight line—horizontally, vertically, or diagonally. Playing this word game is, therefore, the same as playing either tick-tack-toe or the 15 game.

2

HOT	FORM	WOES
TANK	HEAR	WASP
TIED	BRIM	SHIP

See if you can think of other sets of nine words that can be used for this game. The words need not, of course, be in English. Also, why not use sets of symbols, such as the set shown in Figure 3?

3

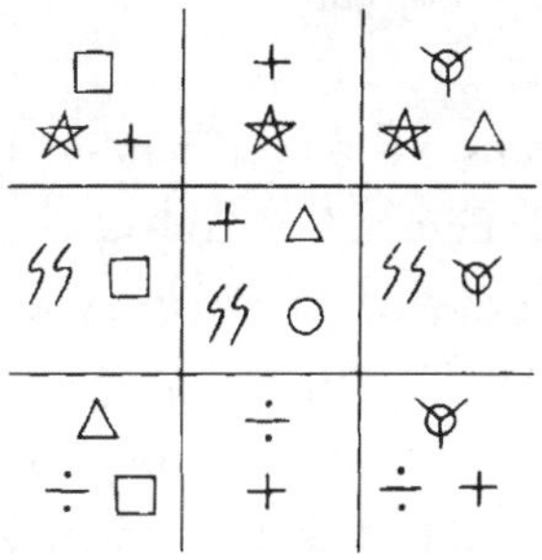

☐ ☆ +	+ ☆	♉ ☆ △
ϟϟ ☐	+ △ ϟϟ ○	ϟϟ ♉
△ ÷ ☐	÷ +	♉ ÷ +

The best way to play all of these games is to write each of the digits, words, or symbols on one of nine blank cards. The cards are spread face up on a table, and two players take turns drawing a card until one player wins.

After you fully understand the isomorphism of all these games, consider the following network game. It is played on the road map shown in Figure 4.

Eight towns are connected by roads. One player has a pencil of one color, the other player has a pencil of another color. They take turns coloring the complete length of any road. Note that some roads pass through towns. If this is the case, the entire road must be colored. The first to color three roads that enter the same town is the winner. At first glance this game seems to have no relation whatever to the games we have analyzed. Actually, it too is isomorphic with tick-tack-toe!

4

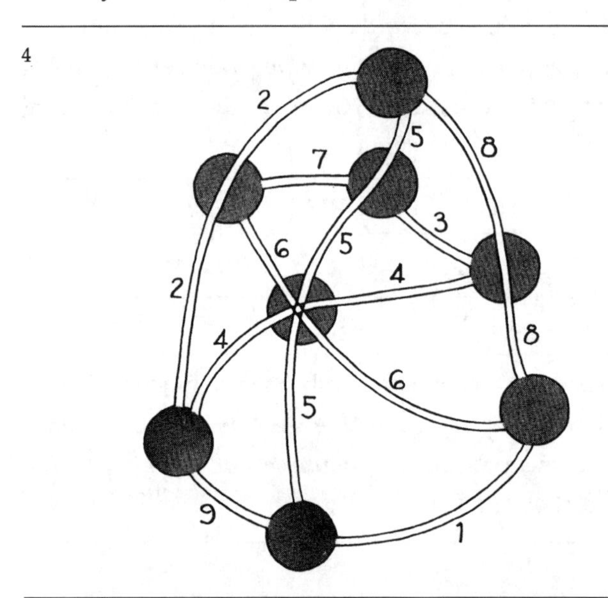

The isomorphism is established by numbering the roads as shown in Figure 4. Each row corresponds to a numbered cell on the magic square. Each town on the map corresponds to a straight line of three cells on the magic square. As before, the isomorphism is complete. Anyone who plays a perfect game of tick-tack-toe can also play a perfect map coloring game.

Figure 5 shows one of the 880 different kinds (not counting rotations and reflections) of 4-by-4 magic squares. The magic sum is 34. Would such a square provide a key for playing a perfect game of 34? That is, a game in which players alternately choose a number from 1 through 16 (no number can be chosen twice) until a player wins by getting four numbers that add to 34. Is this game isomorphic to a 4-by-4 game of tick-tack-toe played on the magic square shown? The answer is no. Do you see why?

Is it possible to alter the rules of tick-tack-toe, allowing winning four-cell patterns other than straight lines, so that an isomorphism can be established between the two games?

5

7	12	1	14
2	13	8	11
16	3	10	5
9	6	15	4

4×4 幻方与 3×3 幻方大不相同，如果不考虑反射和旋转，3×3 幻方只有一种，而 4×4 幻方有 880 种。

Hippo Hangup

关于河马的难题

Many years ago the chief of a wealthy tribe took very good care of the tribe's sacred hippopotamus.

这个问题就是中国古代的曹冲称象问题。

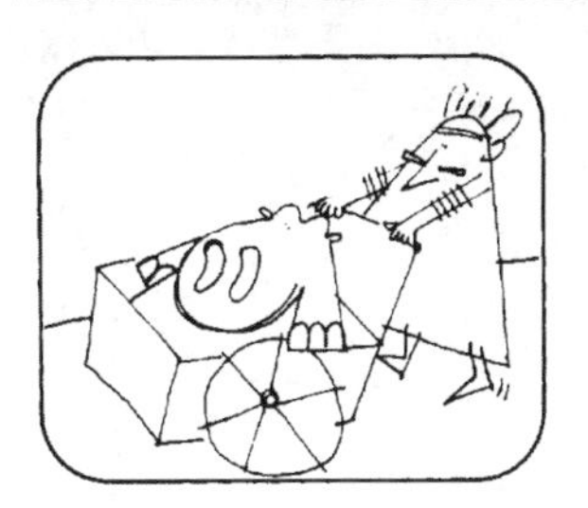

Every year, on the chief's birthday, he and his tax collector would take the animal with them in the royal barge for a trip up the river to the tax collection hut.

It was the natives' custom to give the chief the number of gold pieces that were necessary to match the mass of the sacred hippo. Beside the hut was a large balance that could be loaded with the hippo on one side and balanced with gold pieces on the other.

The chief fed the sacred hippo so well, and the hippo grew so much larger, that one year the balance broke. There was no way to repair the beam without several days of delay.

The chief was livid. He told his tax collector.

The Chief: I want my gold today. And it must be the proper amount. If you can't think of a way to figure it out before sunset, I'll have you beheaded.

The poor tax collector was so frightened that he could hardly think.

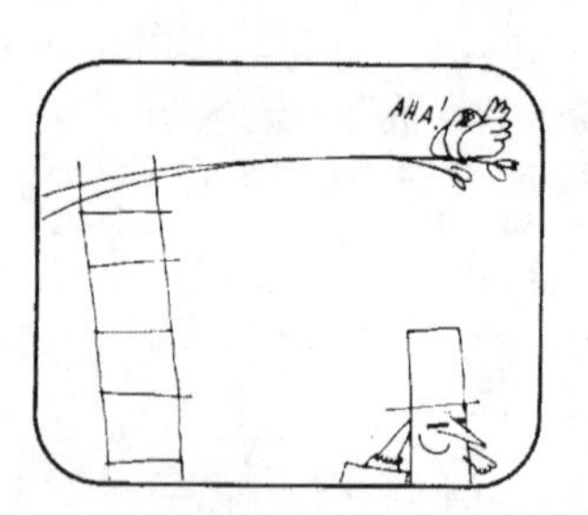

After a few hours of concentration, he suddenly had a brilliant idea. Can you guess what it was?

It was really quite simple. The tax collector put the hippo alone on the royal barge. He marked the water level on the outside of the boat.

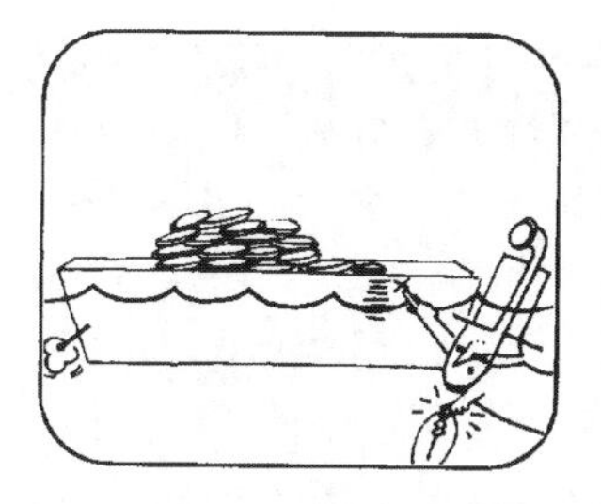

He then had the hippo removed from the barge which was then loaded with gold pieces until the water reached the same mark as before. When that happened, the barge had to contain an amount of gold equal to the hippo's mass.

Eureka!

According to a principle discovered by Archimedes, a floating object always displaces a volume of water equal in mass to the mass of the floating object. Thus, when the hippo is on the barge, the barge sinks deeper into the water, displacing an amount of water with a mass equal to the hippo's mass.

"等量替换"是解决这类问题的主要思路。

Here is a related problem. Suppose that the barge is floating in a tank small enough so that an accurate measure of the tank's water level can be made. The hippo has been replaced by an equivalent mass of gold coins. The level of water on the side of the tank is marked.

Now suppose that all the gold coins are tossed overboard and sink to the bottom of the tank. We know that the water level on the barge will go down. But what about the tank's water level? Will it rise or fall?

Even physicists have difficulty with this question. Some will say that the water level in the tank remains the same. Others will say that it goes up because of the water displaced by the sunken gold. Both answers are wrong.

To see why, let's go back to Archimedes' principle. Every floating object displaces a volume of water with a mass equal to the mass of the object. Gold is much heavier than water, therefore the volume of water it displaces when it is in the barge is much larger than the volume of the gold itself. But when the gold is at the bottom of the tank, it displaces only the amount of water that is equal to its volume. Since this is much less water than it displaces when inside the barge, the water level in the tank must go down.

The physicist George Gamow once explained it in

this dramatic way. Some stars are made of matter that is millions of times as dense as water. A cubic centimeter of this matter would weigh many metric tons. If such a cubic centimeter is tossed overboard and sinks to the bottom, it replaces only one cubic centimeter of water—a trivial amount—therefore, the water level of the tank would go *way* down. The situation with the gold is exactly the same, except that the water level would drop by a much smaller amount.

After the gold has been tossed out of the barge, suppose that a new water level mark is placed on the barge's side. The hippo decides to go for a swim in the tank. When he enters the water, assume that the level in the tank rises by 2 centimeters. How much higher does it rise above the mark on the barge?

Imagine that you are drinking kinky cola from a bottle. You wish to leave in the bottle an amount of cola equal to half the volume of the entire bottle. An easy way to do this is just to drink until the horizontal surface of the liquid in the tilted bottle reaches the spot where the bottle's bottom and sides meet.

Here is a similar problem that must be solved by a different procedure. A transparent glass bottle has an irregular shape. It contains a powerful acid. Only two marks are on the bottle's side. The higher mark indicates 10 liters of acid, the 10wer mark indicates 5 liters.

Someone has used an unknown small quantity of the acid, lowering the liquid's level a trifle below the 10-liter mark. You wish to pour out of the bottle exactly 5 liters to use in an experiment. The acid is too dangerous and volatile to pour into other measuring containers. By what simple procedure can you make sure that you pour out exactly the right amount?

The clever solution is to put into the bottle a quantity of small glass marbles until the level rises to the top mark. Now merely pour out the acid until the level falls to the lower mark.

Dividing the Chores

分配家务

合理分配问题是一个好的数学问题。

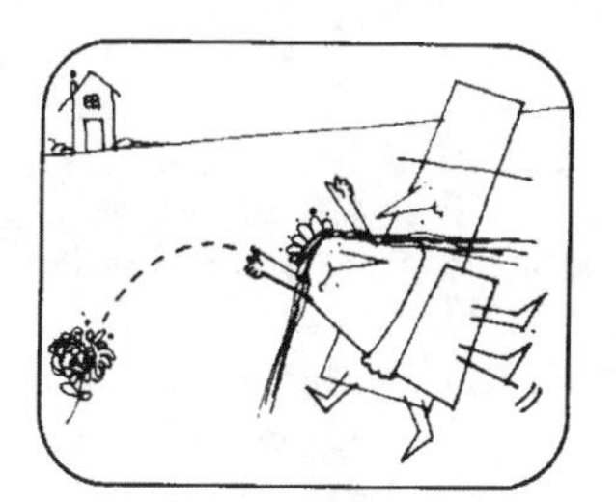

Mr. and Mrs. Buster Jones have just been married. Each has a steady job, so they have agreed to share the household chores.

To divide the chores fairly, the Joneses made a list of all the jobs that had to be done in their apartment every week.

Buster: I've checked half the items, my love. Those are the chores for you to do.

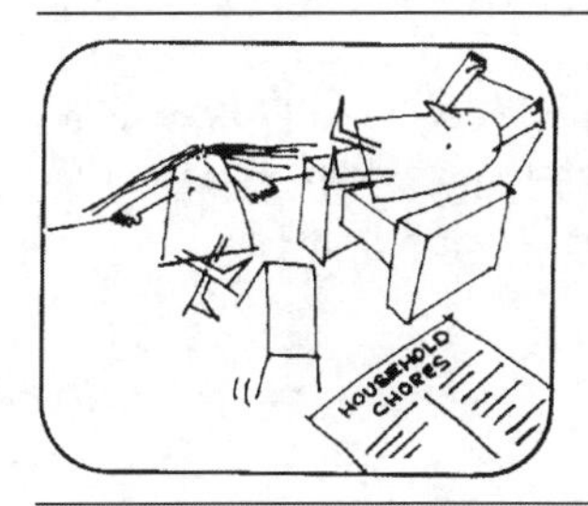

Janet: Sorry Buster, but I don't think you divided the list fairly. You've given me all the dirty jobs and you've taken all the easy ones.

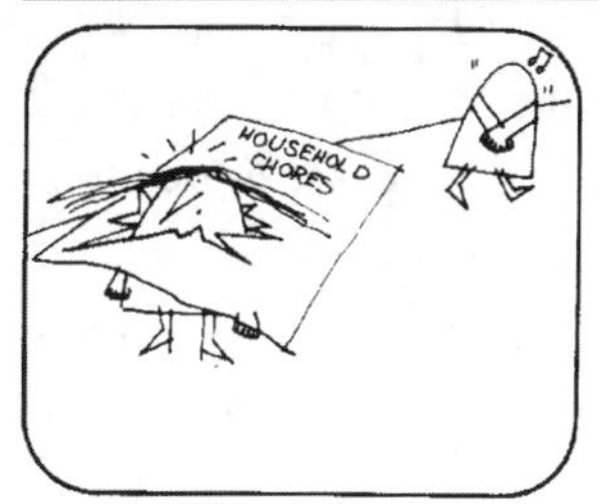

Then Mrs. Jones went over the list and marked the jobs she wanted to do. But Buster wouldn't agree.

Janet: If you expect me to do all these things you're out of your mind.

While they were still arguing the doorbell rang. It was Mrs. Jones' mother.

Mrs. Smith: What are you two love birds fighting about? I could hear you shouting as soon as I got off the elevator.

Mother listened while Buster and his wife explained the trouble. Suddenly she smiled and said.

Mrs Smith: I've just thought of a marvelous solution. I'll show you how to divide the chores so both of you will be completely satisfied.

Mrs. Smith: One of you splits the list into two parts so that you would be willing to take either part. Then the other person gets to pick the half he or she wants first. Isn't that simple?

But it wasn't so simple a year later when Mother moved into the apartment. She agreed to take over one third of the chores but they couldn't decide how to divide the list fairly between the three of them. How do you suggest they do it?

Fair Division

The fair-division problem that has been answered is more usually given in terms of dividing a cake between two people so that each is satisfied he/she has at least half. The problem left unanswered is the same as that of dividing a cake fairly among three people so that each is satisfied he/she has at least a third.

The puzzle of fairly dividing a cake into thirds can be solved as follows. One person moves a large knife slowly over a cake. The cake may be any shape, but the knife must move so that the amount of cake on one side continuously increases from zero to the maximum amount. As soon as any one of the three believe that the knife is in a position to cut a first slice equal to 1/3 of the cake, he/she shouts "Cut! " The cut is made at that instant, and the person who shouted gets the piece. Since he/she is satisfied that he/she got 1/3, he/she drops out of the cutting ritual. In case two or all three shout "Cut! " simultaneously, the piece is given to any one of them.

The remaining two persons are, of course, satisfied that at least 2/3 of the cake remain. The problem is thus reduced to the previous case, and the cake can be fairly divided by having one person cut and the other choose.

This clearly generalizes to n persons. As the knife moves across the cake, the first person to shout "Cut!" gets the first slice (or it is given arbitrarily to one of the two or more who shout simultaneously). Then the procedure is repeated with the $n-1$ persons who remain. This continues until only two persons are left. The final portion of cake is divided as before, or, if you prefer, simply by continuing the procedure with the moving knife. The general solution is an excellent example of proving an algorithm by mathematical induction. It is easy to see how the algorithm can be applied to a list of household chores to be divided among n participants so that each person is satisfied he is getting his fair share.

John H. Conway, a Cambridge University mathematician, has investigated the fair-division problem when the satisfaction demanded by the participants is much stronger. Instead of a procedure that gives each person what he/she thinks is at least his/her fair share, is there a procedure which guarantees that each person is also convinced that no one else has a share greater than his/her own? If you think about it you will see that the algorithm given above does not provide this guarantee if there are three or more people. Conway and others have found solutions for this stronger version when there are three participants, but so far as we know the problem remains unsolved for four or more persons.

The Crooked Acrobat

杂技扒手

这是典型的程序问题。

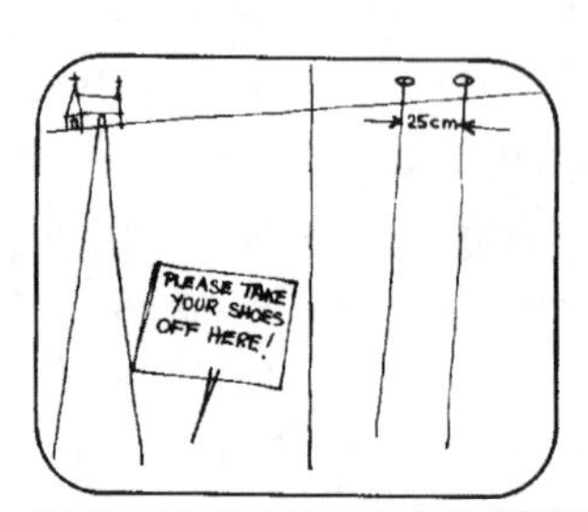

In the tower of a medieval church there are two priceless old bellropes. They pass through small holes in a high ceiling. The two holes are 25 centimeters apart and just big enough to allow the ropes through.

Tony is an acrobat who became a thief. He wants to steal as much of both ropes as he can by cutting them with a knife.

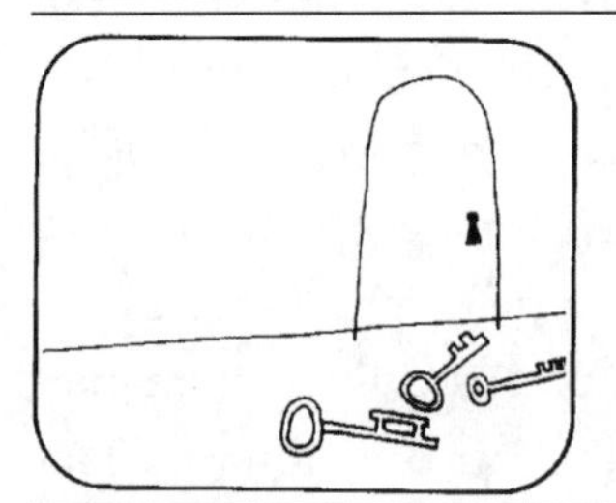

Tony: How shall I do it? I can't get to the room above because the door is triple locked.

Tony: I'll have to climb the ropes and cut them as high as I can. But the ceiling is so high that if I cut them off one-third of the way up I can't drop to the floor without breaking my legs.

Tony thought about it a long time before he hit on a way to get almost all of both ropes. What would you do if you were he?

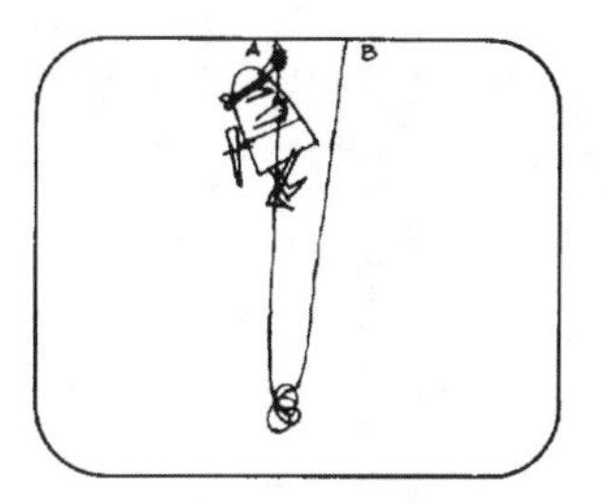

Tony's solution was really clever. First of all he tied together the bottom ends of both ropes. Then he climbed to the top of one rope. Suppose it's rope A.

When he was at the top he cut rope B about half a meter below the ceiling. He tied what was left of B into a loop.

Then Tony put one arm through the loop and hung there while he cut rope A near the ceiling, taking extreme care not to let it fall. Next he passed rope A through the loop and pulled it until the knotted ends were at the top.

He was now able to climb down the double rope, pull it free of the loop, and make off with all of rope A and almost all of B. Could you have done that?

Rope and Other Tricks

Because this story problem is not sharply defined, it has more than one solution. The one just given is probably the most practical, but you may be able to think of many other procedures that the thief could have used. Some may be even better than the solution given here.

For example, the thief could tie a sheepshank knot at the top of rope *B*, as shown in Figure 6. Hanging on *B*, he cuts *A* at the top and lets the end fall. Then he cuts the

6

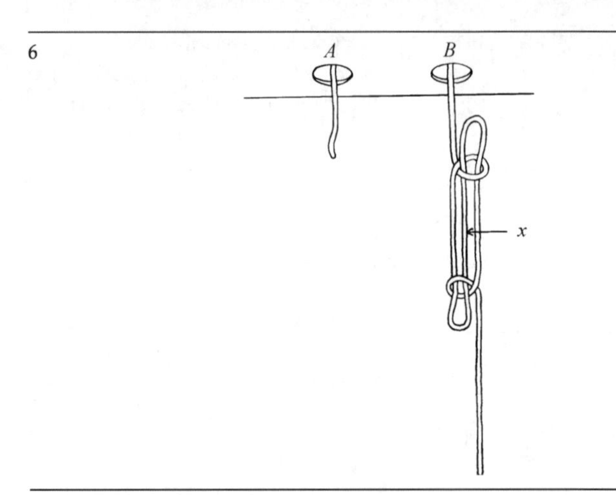

middle strand of the sheepshank at spot X. As all mountain climbers know, the knot will hold until he slides to the ground on rope *B*. Shaking rope *B* will release the knot, giving the thief all of *B* except a small portion at the top.

Another possibility: The thief climbs to the top of rope *A*. With one hand he grabs *B*, keeping his weight on *A*, then with his knife he weakens *A* at the top until he feels the rope is about to break. He then brings *A* and *B* together and hangs by both ropes while he frays *B* at the top in the same manner that he cut *A*. The two damaged ropes support his weight while he goes down both of them. At the bottom, a hard yank on each rope snaps them off at the top.

A third method assumes that the holes in the ceiling are fairly large. First the thief ties *A* and *B* together at the bottom. He climbs *A*, cuts *B* at the very top, then pushes its end up through its hole. By reaching up into the hole for *A*, he grasps the end of *B* and pulls it down through the hole until the cut end is near the floor and the knot is near the top of the hole for *B*. He now can grasp and hold together the top of rope *B* and what was formerly the bottom of rope *A* but now is near the ceiling under the hole for *B*. While he hangs on this double rope he cuts rope *A* off at the top, directly beneath its hole, then slides down the double rope, and finally pulls the rope down.

Here is a clever variation on the previous method. The ropes remain untied at the bottom. The thief climbs rope

A, cuts *B*, pushes the end up through the hole and down through the other hole. The end is then looped around itself and firmly knotted as shown in Figure 7. The thief transfers to rope *B*, cuts *A*, and ties the end of *A* to the knot. He goes down *B*. Now he has only to pull on *A*. Rope *B* slides through its own loop, and both ropes are pulled free of the ceiling.

7

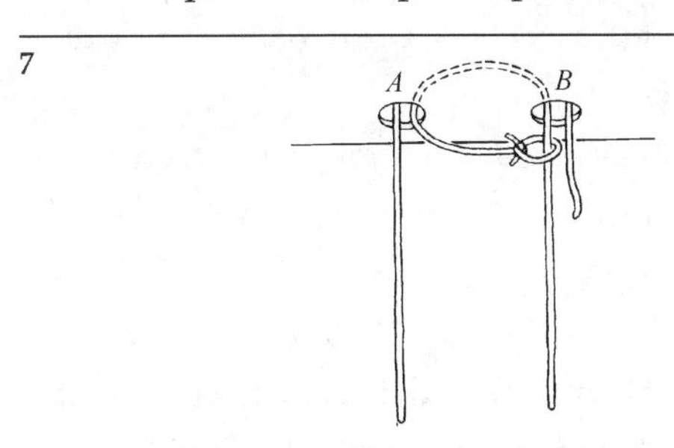

Still another variant. The thief climbs *A*, ties a loop at the top of *B*. He hangs on this loop, cuts *A*, brings its end up through the hole and down through the hole for *B*. He ties the end to the loop. Hanging on both ropes, he cuts *B* at the top, above the loop, goes down both ropes, then pulls on rope *B* to obtain all of both ropes.

Some of these methods undoubtedly would cause the bells to ring and the thief to be caught. One of the virtues of the original solution is that the thief, by pulling gently on rope *B* before he hangs in its loop, can avoid ringing bell *B*. Of course, when he first climbs *A* he also pulls on it slowly before he starts his climb.

A number of classic procedural puzzles, similar to river crossing problems, involve the use of a long rope that goes over a pulley and has a basket at each end. Here is one version of such a problem that was a favorite of Lewis Carroll.

A queen and her son and daughter are being held captive in the top room of a high tower. Outside their window is a pulley with a rope over it, and a basket at each end of the rope. The baskets are of equal weight. The one outside the window is empty, and the one on the ground contains a stone with a mass of 30 kilograms. The stone serves as a counterweight.

Lewis Carroll 以他《艾丽丝漫游奇境》和《镜中世界》两部最受欢迎的儿童读物著称于世。但他是学数学出身，长期在牛津大学担任数学讲师，对数学也有研究。他的这个问题是最好的程序数学难题之一。

There is enough friction in the pulley so that it is safe for anyone to be lowered in one basket provided his or her mass is not greater than the mass in the other basket by more than 6 kilograms. If the difference is greater than 6 kilograms, they come down with such speed that the bump at the bottom might injure them. Of course, when one basket goes down, the other basket goes up to the window.

The queen's mass is 78 kilograms, the daughter's is 42 kilograms, and the son's is 36 kilograms. What is the simplest algorithm—simple in the sense of the fewest number of steps—by which they all can get safely to the ground? The basket is large enough to hold any two people, or one person and the stone. No one assists the prisoners in escaping, nor can they help themselves by pulling on the rope. In other words, the pulley operates only when the mass in one basket exceeds the mass in the other.

The simplest solution is most easily found by simulating the problem. Write the masses on separate cards, and move the cards up and down. You will not be able to get all three persons down in fewer than nine steps. Here's how it's done:

1. Son down, stone up.
2. Daughter down, son up.
3. Stone down.
4. Queen down, stone and daughter up.
5. Stone down.
6. Son down, stone up.
7. Stone down.
8. Daughter down, son up.
9. Son down, stone up.

Problems of this sort are sometimes made more difficult by including animals that cannot climb in and out of the baskets without the help of persons. Lewis Carroll proposed the following version of the preceding problem. In addition to the Queen, son, daughter and stone there are at the top of the tower a pet pig of mass 24 kilograms, a dog of 18 kilograms, and a cat of 12 kilograms. The restrictions on differences in mass between the two baskets remain the same, but now there must always be someone at each end to put the pets in and out of the baskets.

See if you can find a solution in as few as 12 steps. Note that in both problems the last person to step out of the basket must move aside quickly, otherwise the basket with the stone may drop on his/her head!

Island Crackup

飞机坠落于小岛

这个问题显示数学在解决实际问题中的作用，通过抽象及逻辑代替盲目实践。

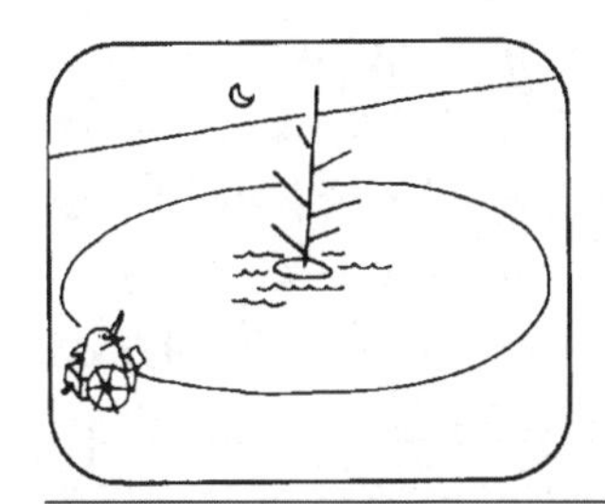

Orville has parked his car at the edge of a small lake.

Orville: This is a nice flat spot for flying my radio controlled model plane. There are no trees or rocks except that one big tree on the little island in the middle of the lake.

Orville tried to maneuver his plane around the tree, but he misjudged the distance and the plane crashed into the tree and fell to the ground.

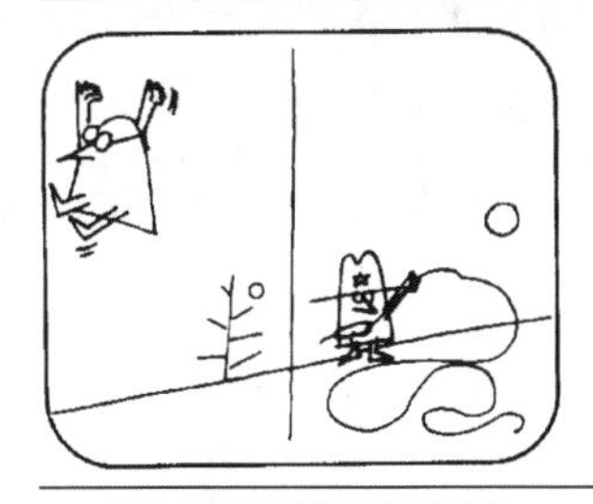

Orville was very upset. He wanted to get his expensive plane back, but the water was very deep and he didn't know how to swim. He had a rope in his car that was a few meters longer than the widest part of the lake but he didn't see any way to use it.

All at once Orville had a flash of insight.

Orville: Why there's nothing to it. I'll get wet, but I'll have my plane back in no time.

What procedure did Orville think of?

数学除了逻辑之外还有计算，在思考时往往还要考虑优化问题。

Thinking Instead of Swimming

Orville retrieves his model plane by the following ingenious procedure. He ties one end of the long rope to the front bumper of his car, which is parked close to the edge of the lake. Holding the tree end of the rope he walks completely around the tree at the island's center. He tightens the rope and ties the free end to the bumper. This creates a firm double rope stretched between car and tree. Although Orville can't swim, he can pull himself through the water, hand over hand along the double rope, and easily get over to the island and back.

Another splendid old puzzle also deals with the task of using material on hand for getting from a shore to a small island. In this case the "island" is in the center of a square lake as shown in Figure 8. A man wishes to get from the shore to the island. As before, he cannot swim. Two identical planks are on the shore, but each plank's length is a trifle too short to extend from the side of the lake to the island. How does he manage to use the two planks for getting to the island?

Figure 9 shows the solution.

Let us generalize by assuming there are more than two identical planks. Can the planks be shorter than those used before, and still produce a bridge to the island?

You may have little difficulty thinking of the 3-plank bridge shown in Figure 10, but not many persons are likely to discover how five or eight planks can be still shorter and still span the water. Figure 11 shows the 8-plank solution.

We can idealize the problem by making the island a point and the planks line segments, allowing them to "overlap" by just touching one another. Imagine the procedure extended to an infinite number of identical planks. The limiting case is shown in Figure 12. If the side of the square lake is 2 units, then $\sqrt{2}/2$ is the shortest length each plank can be, provided there is an infinite number of them. This can be proved by applying the Pythagorean theorem.

You may enjoy investigating similar idealized plank problems for "lakes" with boundaries other than squares, such as circles and regular polygons.

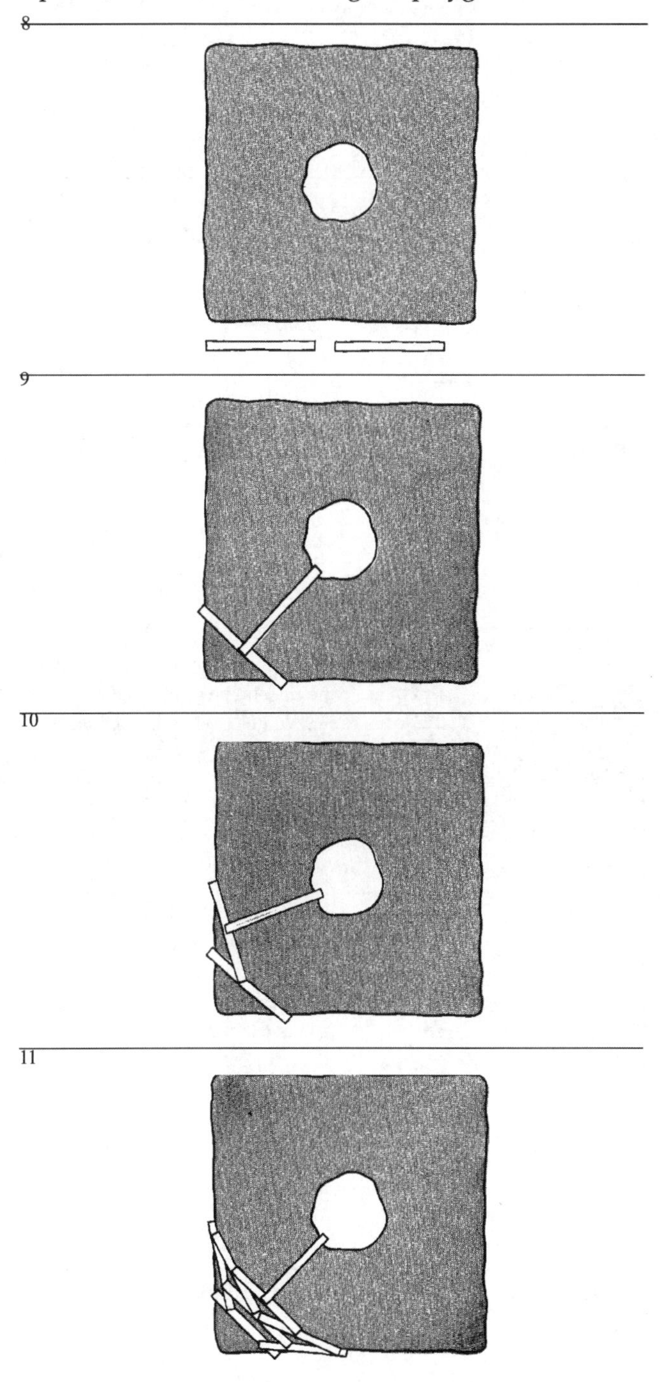

12

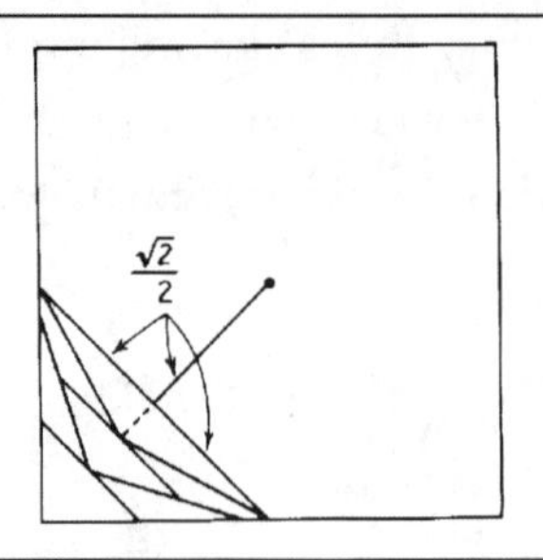

The Lazy Lover

懒惰的朋友

Jack fancies himself as the world's greatest lover. He is planning to rent an apartment in Washington D. C.

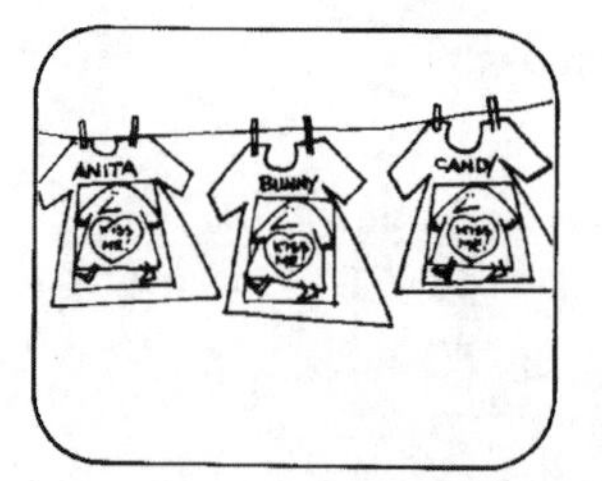

Jack has three girl friends who live in the city and he wants to live in a spot that will be as close as possible to all three.

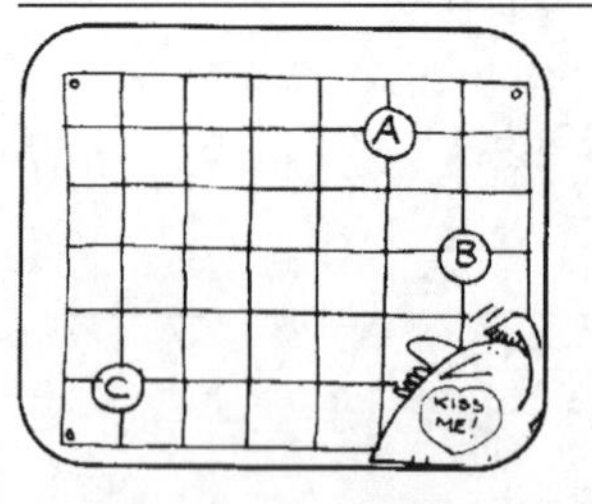

Jack marked the corners where the girls lived on a city map.

Jack: Let's see now. I must pick a spot to live so that the sum of the distances to each girl's house is as small as possible.

Jack tried and tried, and was about to give up when he shouted.

Jack: Aha! I see an easy way to find the spot I'll live in.

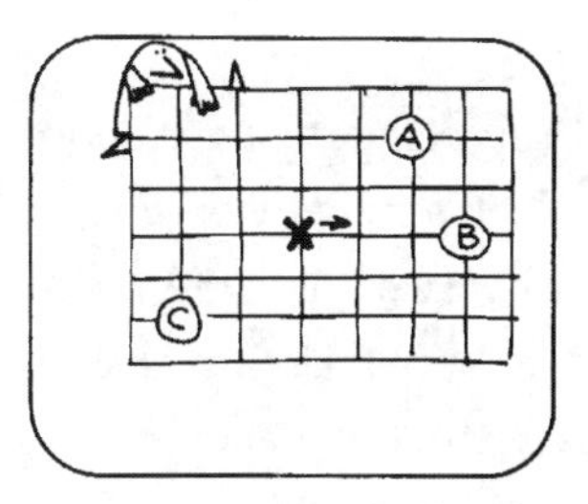

Jack's clever procedure was to ask himself how the girls would vote if he moved from one place to another. He started with a spot that looked reasonable and then considered moving a block east.

Jack: Anita and Bunny would vote "yes" for this move because I'd be close to them. Candy would vote "no ". But the distance I save is more than I lose, so I'll accept the majority vote.

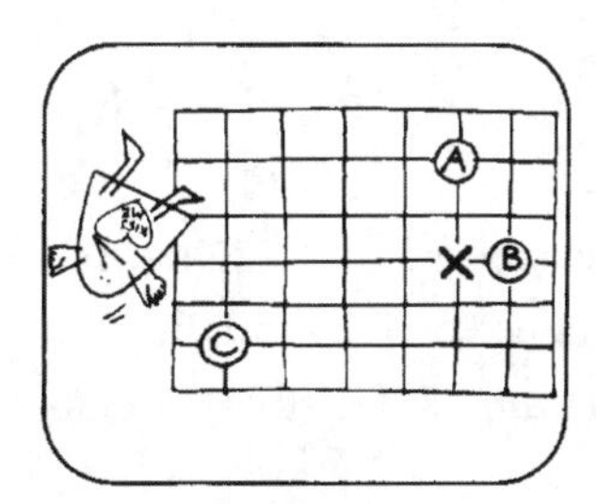

So whenever a majority vote was "yes", Jack made the move. And whenever it was "no", he tried a different move. Eventually he reached a corner where he couldn't move without a "no" vote. This was where he decided to live.

Luckily, Jack was able to rent an apartment at just the right place. Then, a week later, Bunny moved 7 blocks away.

Jack: Holy cow! Now I'll have to move to a new location. But when Jack checked the map he was surprised to find that he could stay right where he was. Can you explain how this could happen?

这是一个重要的运筹学问题。

Voting Algorithm

If Bunny moves seven blocks due east, her new residence has no effect on the location of Jack's residence. Indeed, she could move any distance whatever east, and Jack's present apartment will continue to be at the optimal spot.

You can better appreciate the efficiency of the voting algorithm if you try it on larger grids on which more than three spots are marked. You will find that the procedure quickly locates the position x that minimizes the sum of the distances for x to all spots, but only provided the number of spots is odd. Why does it not work when the number is even? The answer is that if the number is even a tie vote is possible. Whenever a tie vote occurs, the procedure halts.

You may wish to investigate the following related questions:

1. Can you invent a procedure that applies when the number of spots is even?

2. Under what conditions can the displacement of one or more spots have no influence on the location of x?

3. Is the voting procedure affected if street widths are taken into account?

4. Is it affected if the spots, including x, are not confined to street intersections?

5. Will the voting procedure work if the grid is composed of straight streets that may have any orientation on the plane?

6. Will the procedure work if streets are crooked or curved?

Although the voting procedure applies to any sort of network, it fails on the unmarked plane because travel is no longer confined to certain paths. The general problem is this. Given n points on the plane, find a point x such that the sum of the straight-line distances from x to all the points is as small as possible. For example, consider three cities, A, B and C. Where should an airport be located so that the sum of the distances by plane to the three cities is minimal? This obviously is not the same as asking for a minimal sum of distances to the cities by car. In other words, the ideal location for an airport may not be the same as an ideal location for a bus station.

The answer, which is not easy to prove geometrically, is that the airport should be where the three lines from it to the three cities make three 120°-angles. In the case of four cities, if they are the corners of a convex quadrilateral, the airport should be at the intersection of the two diagonals. This is not hard to prove. The general problem of locating x for any number of given points on the plane is more difficult.

Can you think of a simple mechanical device (analog computer) that quickly finds the location of x for any three points on the plane? Let the plane be represented by the surface of a table. At each of the three spots we drill a hole through the table top. Tie the ends of three pieces of string together. Pass the free ends through the holes, one to a hole, and attach weights of equal mass to each end. The equal forces on the strings correspond to the three equal "votes" by residents at the three spots. The position assumed by the knot above the table indicates spot x. This works, of course, because of an isomorphism between the problem's mathematical structure and the structure of the physical model.

Let us now complicate our original puzzle. Suppose that instead of a single girl friend at spots A, B, C, these spots represent buildings in which school children are living. There are 20 children at A, 30 at B, and 40 at C. All attend the same school. Where should the school be located so as to minimize the sum of the walking distances of all 90 pupils?

If their walking is restricted to the streets of a city, we can apply the same voting procedure as before, allowing each child a single vote. This will soon find the spot where the school should be located. However, if the three buildings are on a plane, and the students may walk to school in direct straight lines (such as children in the country who can cut across open fields), can we modify our analog computer so it works as well as before?

Yes. Instead of equal weights, we use unequal weights with masses that are proportional to the number of pupils in each building. The strings will assume a position at which the knot locates the school.

Will our computer work if the number of pupils at one building is more than the sum of the pupils at the other two? For example: 20 children at A, 30 at B, and 100 at C? Yes, it works just as well. The weight corresponding to the 100 students will pull the strings until the knot catches at the top of hole C. This indicates (correctly!) that the school should be at site C.

Will our analog computer work for more than three spots? Yes, it generalizes to n spots even when they are not corners of a convex polygon. However, friction becomes such a factor that with more than three spots the system does not work efficiently.

Graph theory is a rapidly growing branch of mathematics concerned with vertices (points) connected by lines. There are dozens of important graph theory applications to the finding of minimal paths. Some have been solved, some remain unsolved. An example of a famous solved problem is the following.

Given n points on a plane, join them to one another by straight lines so that the total length of the network is as short as possible. We are not allowed to add new vertices to the plane. Such a network is called a "minimal spanning tree". Can you invent an algorithm for finding such a network?

Kruskal's algorithm (named after Joseph B. Kruskal who was the first to give it) finds the minimum network as follows:

Determine the distances between every pair of points, and label these distances in increasing order of lengths. The shortest is 1, the next shorter is 2, and so on. If two distances are equal it does not matter which is numbered first. Draw a straight line between the two points separated by distance 1. Follow with straight lines for distances 2, 3, 4, 5, and so on. Never add a line that completes a circuit. If drawing a line produces a circuit, ignore that pair of points and go on to the next higher distance. The final result is a minimal spanning tree connecting all points.

Such spanning trees have interesting properties. For example, the lines will intersect only at the spots, and no more than five lines will meet at any point.

Minimal spanning trees are not necessarily the shortest networks joining n points. Remember: We are restricting the network to one that does not have additional vertices. If new vertices are allowed, the network may be shorter. A simple example is provided by four corners of a unit square. The minimal spanning tree consists of any three sides of the square (Figure 13 left). Suppose we are allowed to add new vertices. Is there a network joining the four corners that is shorter than 3?

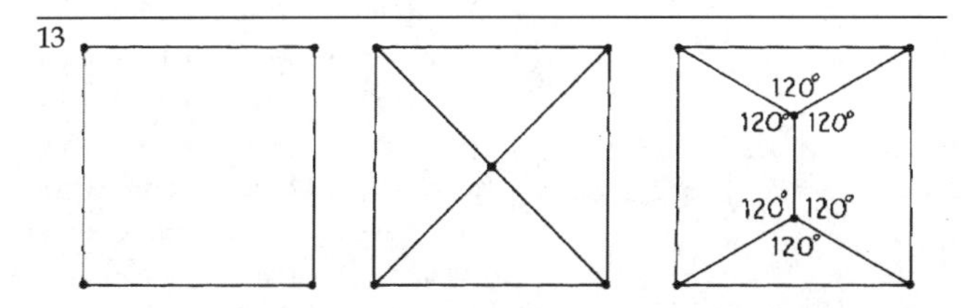

Most people assume that the minimal network consists of the square's two diagonals (Figure 13 center), but this is not the case. Figure 13 right shows the solution. The two diagonals of the square have a total length of $2\sqrt{2}=2.82^{+}$. The network in Figure 13 has a shorthen total length of $1+\sqrt{3}=2.73^{+}$.

The general problem of finding a minimal length network connecting n points on the plane, when new vertices are permitted, is known as Steiner's problem. It is solved in special cases, but there is no known efficient algorithm that locates the "Steiner points" (new vertices) of a minimal Steiner tree that connects n points on the plane. The problem has many engineering applications, from the design of microprocessor chips used in electronic calculators to the finding of minimal networks for railroads, airplane routes, telephone lines and other forms of travel and communication.

Sanitary Surgeons
外科医生

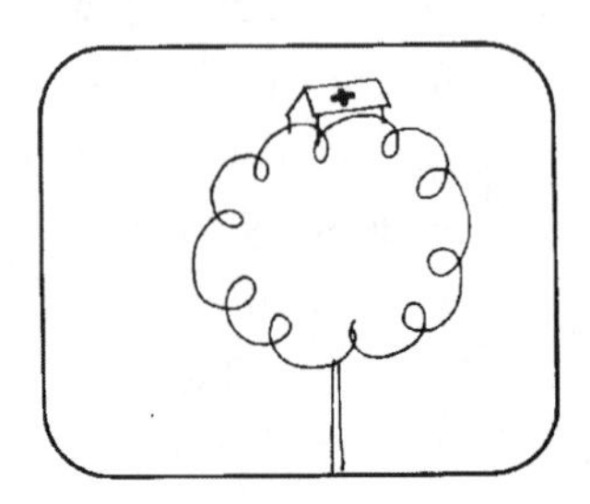

Deep in a tropical jungle there is a hospital where three surgeons—Jones, Smith, and Robison—are on staff.

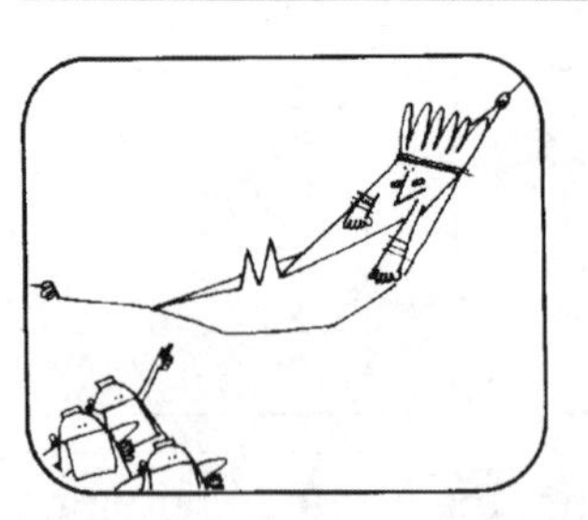

The local tribal chief is suspected of having a rare disease that is highly infectious. The three surgeons must operate on him, one at a time. To complicate matters any one of the three doctors could have caught the disease while examining the chief.

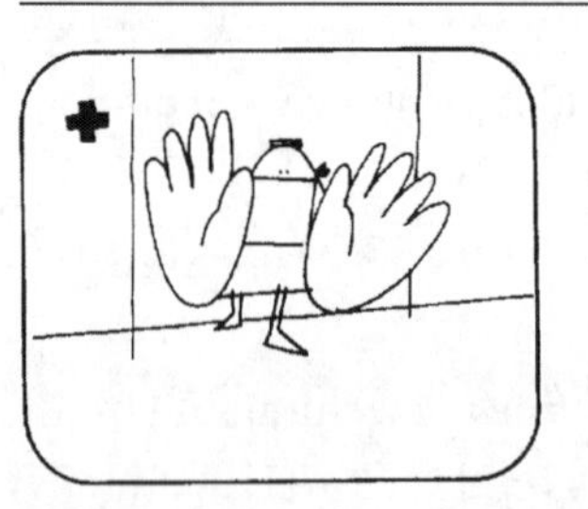

Each surgeon must wear rubber gloves when he operates. If he has the disease, its germs will infect the side of any glove that touches his hand. And if the chief has the disease, it will contaminate the outside of any glove worn.

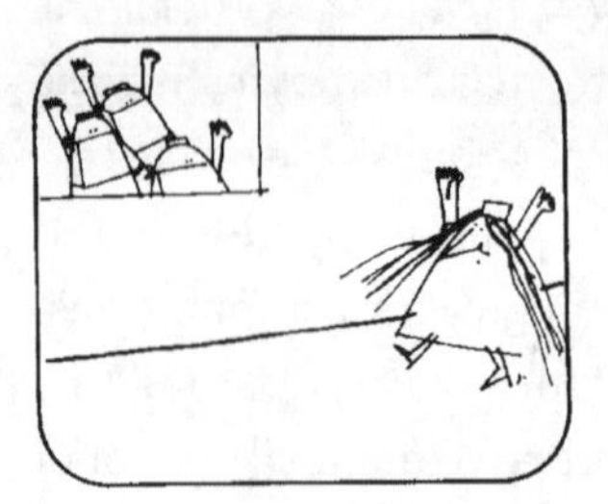

The operation was about to start when Miss Kleene, the nurse, rushed into the room.

Nurse Kleene: I have bad news for you, doctors.

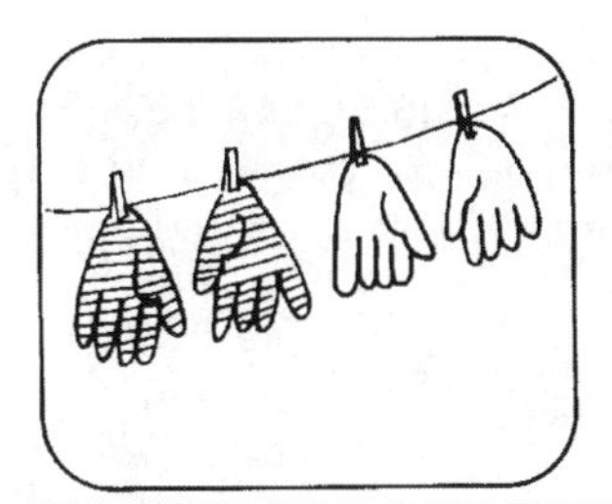

Nurse Kleene: We have only two pairs of sterile gloves left. One pair is blue, and the other pair is white.

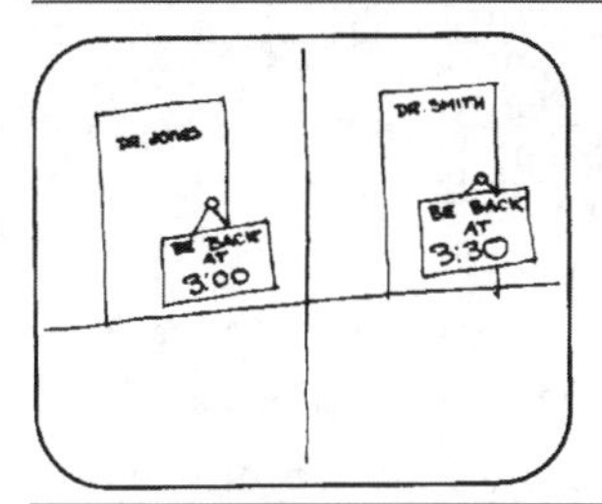

Dr. Jones: Only two pairs? If I operate first both sides of my gloves could be contaminated. If Smith operates next both sides of his gloves could be contaminated. Then no sterile gloves will be left for Robison.

Suddenly Dr. Smith made a suggestion.

Dr. Smith: Suppose I wear both pairs of gloves, the blue on top of the white. One side of each pair might get infected, but each pair would still have one side that was sterile.

Jones caught on quickly.

Dr. Jones: I see. I can wear the blue pair, sterile side in. Then Robison can reverse the white pair and wear them sterile side in as well. None of us will run the risk of catching the disease from the chief or from each other.

Nurse Kleene: That's fine for you doctors, but how about the chief? If any of you is infected, and the chief isn't, he could catch the disease from one of you.

The surgeons were floored by this remark. What should they do? A moment later Miss Kleene exclaimed.

Nurse Kleene: I know how all three of you can operate without you or the chief running any risk of catching the disease.

None of the doctors could figure out what Miss Kleene had in mind. But when she explained, they all agreed that it would work. Can you figure it out?

---

blue
red
green

Inside and Out

Before explaining Miss Kleene's brilliant procedure, let's make sure we understand the first procedure that protects only the surgeons.

Let *W*1 stand for the insides of the white pair of gloves, and *W* 2 for the outsides. Let *B*1 stand for the insides of the blue pair, *B*2 for the outsides.

Dr. Smith puts on both pairs of gloves, first the white, then the blue. Sides *W*1 can become contaminated by him, and sides *B*2 may become contaminated by the tribal chief. After Smith operates he removes both pairs of gloves. Dr. Jones puts on the blue pair with sterile sides *B*1 touching his hands. Dr. Robison then turns the white pair inside out and wears them with sterile sides *W* 2 touching his hands.

Now for Miss Kleene's procedure.

Dr. Smith wears both pairs of gloves as before. Sides *W*1 and *B*2 may become contaminated, while *W*2 and *B*1 remain sterile.

Dr. Jones wears the blue pair with sides *B*1 against his hands.

Dr. Robison turns the white pair inside out and puts them on with sides *W*2 against his hands. Then he puts the blue pair on, over the white, with sides *B*2 on the outside.

In all three cases only sides *B*2 touch the chief. therefore he runs no risk of catching the disease from any of the surgeons.

So far as we know, this problem has not yet been fully generalized. Given n surgeons who have to operate on k patients, what is the minimum number of pairs of gloves that will guarantee that neither they nor the patients run a risk of catching the disease from one of the others?

Chapter 6

Word aha!

文 字 游 戏

Puzzles about letters, words and sentences

Mathematicians tend to be addicted to word play. For example, there is a famous footnote in a sober textbook *Graphical Enumeration* (by Frank Harary and Edgar M. Palmer) pointing out that "Read and Wright are both wrong;" the reference is to a claim by mathematicians Ronald C. Read and E. M. Wright. We could fill a book with other examples.

It is not hard to understand why mathematicians enjoy such jokes. Words are nothing more than combinations of letters, in an accepted order, just as sentences are strings of words that are linked together according to the formation rules of syntax. Language, therefore, has about it a strong flavor of combinatorial mathematics, with many striking resemblances to combinatorial number theory. Word squares are similar to magic number squares. The use of punctuation marks in a sentence corresponds to the use of mathematical symbols (parentheses, plus and minus signs, and so on) to "punctuate" a sentence of algebra.

All of the above pleasant analogies are examined in this section, as well as many others. The palindrome—a sentence that reads the same backward as forward—is similar to a palindromic number. As we shall see, there is a notorious "palindrome conjecture" in number theory that is still not settled. And there are interesting theorems about palindromic primes and palindromic numbers that are squares and cubes. Other puzzles in this section involve the dividing of words into parts in much the same way that sums are divided into parts in partition theory, a branch of number theory.

If we view letters as geometric figures, a host of unusual puzzles arise. We will see how some problems of this type concern two important kinds of symmetry: 180-degree rotation symmetry (sometimes called "two-fold symmetry"), and mirror reflection symmetry. We will discover that certain words, even entire sentences, can be turned upside down without altering their pattern. The fact

虽然数学家喜欢文字游戏，但是文字游戏与数学游戏关系并不太紧密。因此本章与前面5章不同，前5章的50节都是大大小小的趣味数学难题，而本章的15节中只有两三节真正与数学有关，特别是“标点与符号”（227页）。其他部分主要是文字游戏，要求读者比较熟悉基础英语。最典型的文字游戏是回文与纵横字谜，解决这些难题主要靠灵机一动，不像解数学难题那样可推广到一般情形。
最后两段表示语言游戏最基本的要素之一：一个词既可以代表该词本意，也可以表示该词本身。

that each digit resembles a letter when rotated 180 degrees is the basis for a type of amusement that has become very popular since pocket calculators became common.

Let us view the letters not as rigid patterns that keep their shapes when rotated and reflected, but as topological figures that can be twisted and deformed like elastic strings. This, too, leads to recreational problems that you will find explored here, and which give basic insights into topological structure.

Finally, we will encounter word problems that introduce some important conceptions of mathematical logic. A trivial riddle about the opposite of "not in" ties in with rules about negation in logic, and the handling of negative signs in algebra. Many of our ridiculous riddles become clear only when you recognize that you cannot talk about the words and sentences of a language without expressing yourself in a higher-level language that logicians call a "metalanguage".

We intended this final section of the book to be the lightest and the funniest. Have you wondered why a section on word play appears at all in a book about mathematical recreations? We have already given the answer. It is not just that mathematicians love word play, or that it has a combinatorial aspect, but the fact that even word puzzles can lead into unexpected and significant aspects of serious mathematics.

Dr. Wally O. Wordle

W. O. 沃德尔博士

Meet Dr. Wally O. Wordle, a famous mathematician.

Dr. Wordle is the host of *Winning Wordles*, a popular TV game show that he invented. Guests on the show win fabulous prizes when they solve Dr. Wordle's clever word puzzles.

Dr. Wordle: Word play is just like mathematics. The symbols are letters and words. And the rules of spelling and grammar tell us what particular combinations are allowed.

Dr. Wordle: Let me give you a couple of examples. First, what is the opposite of "Not in" ?

Dr. Wordle: And what 11-letter word do all Yale graduates spell incorrectly?

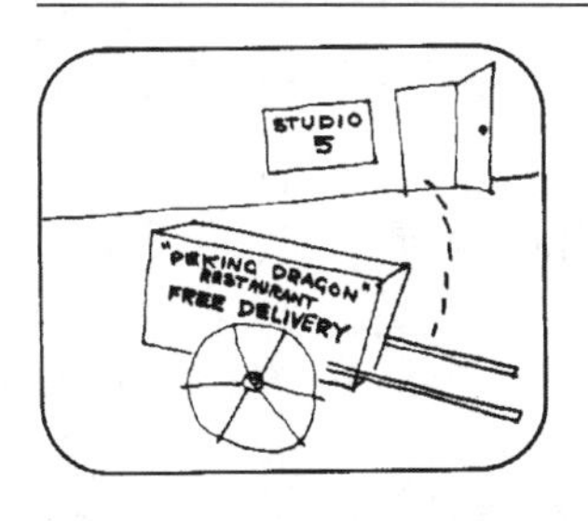

Dr. Wordle: Did you get those two quickies? The opposite of "Not in" is "in". And the word all Yale graduates spell incorrectly is "incorrectly". There's much more of the same on today's exciting show. So let's bring on our first guest.

否定的否定表示肯定，这不仅是逻辑问题，而且也是语言问题。作者幽默地指出在自然语言中，没有肯定的肯定表示否定的情形。但作者没有指出，对于否定疑问句的回答，汉语和多数西方语言的回答正好相反。

Not Not

There is a strong tendency to think that the opposite of "not in" is "out", but, of course, the opposite is "not not-in" which is the same as "in". Two negatives make a positive in grammatically correct English as well as in multiplication and formal logic. In fact, you sometimes encounter a chain of three or more negatives. The rule is that any even number of negatives make a positive, and any odd number of negatives make a negative. Here are some examples of statements that illustrate the rule:

1. $x = (7-3) - [(-4+1)]^3$.

2. Headline in *New York Times* (May 6, 1965) : *Albany Kills Bill to Repeal Law Against Birth Control*.

3. The philosopher Alfred North Whitehead once thanked a speaker for having "left the darkness of his subject unobscured".

4. A young man received the following letter from his girl: "I must explain that I was only joking when I wrote that I didn't mean what I said about reconsidering my decision not to change my mind. I really mean this."

5. Mathematics teacher: "I can't seem to make you understand the meaning of negation, so I'm not going to try any more."

Student: "Ah, I see what you mean, and I'm pleased that you are willing to continue."

6. In the colloquial dialect of certain ethnic groups, double negatives are often used, in violation of the rule, to reinforce a negative. Here are a few examples:

"Don't give me no back talk."

"We aren't never going to stand for nobody using no double negatives around here."

"I got spurs that jingle-jangle jingle,
As I go riding merrily along,
And they sing, 'Oh aren't you glad you're single,'
And that song aren't so very far from wrong."

7. A professor of logic tells his class that he knows of no natural language in which two positives, in violation of the standard rule, are used to mean a negative.

Sarcastic voice from rear of room: "Yeah, yeah."

The riddle about the word "incorrectly" catches people off guard because they take the word to mean an adverb modifying the verb "spell" rather than the word itself. In modern semantics, any question about a word or sentence is in what is called a "metalanguage", while the word or sentence belongs to what is called an "object language". The two languages are frequently distinguished by putting quotation marks around statements in an object language. For example, quotation marks around "incorrectly" in the original wording of the question, would have made the question less ambiguous. Confusion often results by failing to recognize the two levels of language. Here are some puzzling statements that illustrate this:

What-do-you-think was the horse's name.

How Long is a Chinese mathematician.

Can you explain the meaning of this sentence? That that that that that signifies is not the that to which I refer.

And how about this one? Wouldn't the sentence "I want to put a hyphen between the words Fish and And and And and Chips in my Fish-And-Chips sign" have been clearer if quotation marks had been placed before Fish, and between Fish and and, and and and And, and And and and, and and and And, and And and and, and and and Chips, as well as after Chips?

Shee Lee Hoi

西·李·霍

Dr. Wordle picked Mr. Hoi as a guest as soon as he found out his phone number. Do you see the unusual connection between Shee Lee Hoi and his phone number?

这些是数码与字母相似的游戏。

Turn Mr. Hoi's name upside down and it becomes his phone number.

Every electronic numeral can be interpreted as a letter when viewed upside down. This is the basis of scores of stunts with pocket calculators that have become popular in recent years.

The first of these stunts, which apparently started the craze, had a story line about the Arab- Israeli war. The following version was devised by Donald E. Knuth, a well-known computer scientist: 337 Arabs and 337 Israelis were fighting over a square of property that is 8.424 meters on the side. Who won? To find out, obtain the square of 337 and add it to the square of 8.424 to get the sum 71,077.345. When this is looked at upside down it spells "SHELL OIL".

Entire books have been written about numbers that become words when viewed upside down in a calculator's read-out. The following chart shows how each digit, when inverted, resembles a letter:

0	O	5	S
1	I	6	g
2	Z	7	L
3	E	8	B
4	h	9	b

With the aid of this chart, you'll find that it is not hard to make up amusing calculator problems of your own that end with an answer that can be read as an appropriate word when the machine is turned upside-down. If you like, the decimal point can be used to separate two words.

Here are some good examples:

1. What is the capital of Idaho? 4 times 8,777.

2. What did the astronaut say when he first stepped out on the moon? 13,527 divided by 3.

3. The more you take away, the larger it grows. What is it? $\sqrt{13719616}$.

4. If Bourbon whisky is $8 a bottle in Chicago, what is Scotch in New York? 8 times 4,001.

5. What did Dr. Livingstone say after Stanley said, "Dr. Livingstone, I presume?" 18 times 4, then divided by 3 and the result decreased by 10.

6. Are there similar calculator stunts that use foreign words? Add 1 to the previous answer.

Elusive Eight

无从捉摸的"八"(EIGHT)

一个数码可以表为阿拉伯数码(如8),也可以表为字母(如Eight),还可以表为发音类似的词(如ate)。

Dr. Wordle: Mr. Hoi, our first problem uses these 18 chopsticks. For $ 5 can you remove just 8 sticks and leave "8"?

Mr. Hoi: Confucius say, "A problem that cannot be solved should be kissed and left."

Dr. Wordle: Don't give up yet Mr. Hoi. Remember, this is a word game show. "Eight" is a word that can be spelled.

Mr. Hoi: I thought of that. But "eight" has too many letters. There't no way to spell it.

Dr. Wordle: Time's up. Too bad you forgot that "eight" can be spelled another way.

Arithmetic Puns

通过筷子或火柴可以摆出许多数学公式。

The solution to the chopstick puzzle required the insight that the word which Professor Wordle pronounced like the name for the numeral "8" can also be taken as the word "ate".

Here is a variant of the same puzzle that calls for a different aha! The chopsticks (or matches) are arranged the same as before, but this time the task is to take away 13 sticks and leave 8. The solution is to leave the *numeral* 8 as shown below:

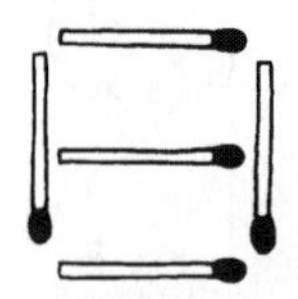

If your friends find the two previous puzzles too easy, try this more difficult version on them. The starting pattern again is the same. Take away 7 sticks and leave 8. The solution this time is to form an expression that equals "8".

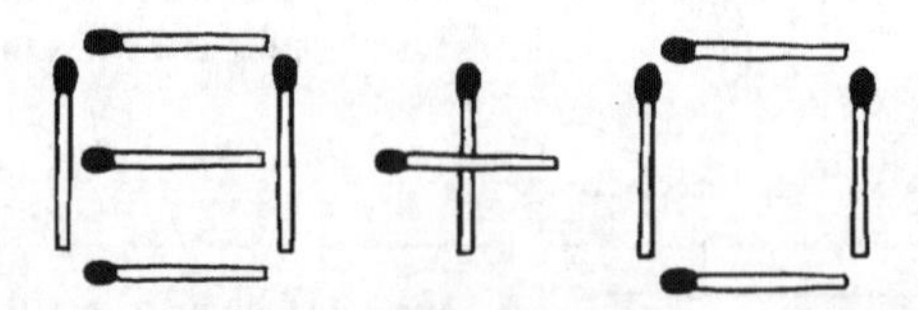

There are endless other puzzles with sticks such as chopsticks, matches, toothpicks, coffee stirrers, soda straws, pencils, or whatever is most convenient. Here are two more to try on friends.

Arrange 12 sticks like this:

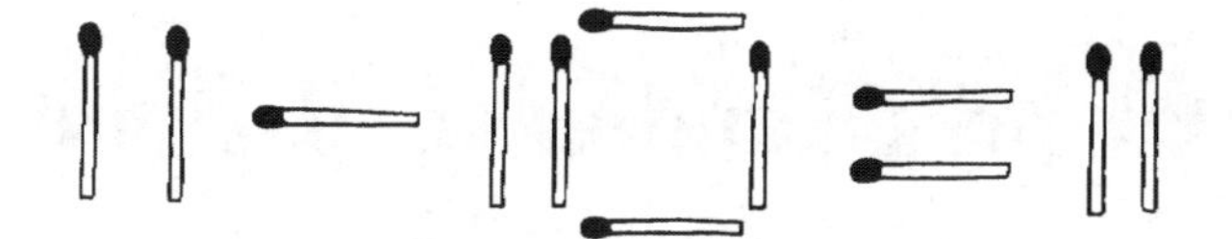

The task is to move one stick to make a correct equation. Here are four of many different solutions:

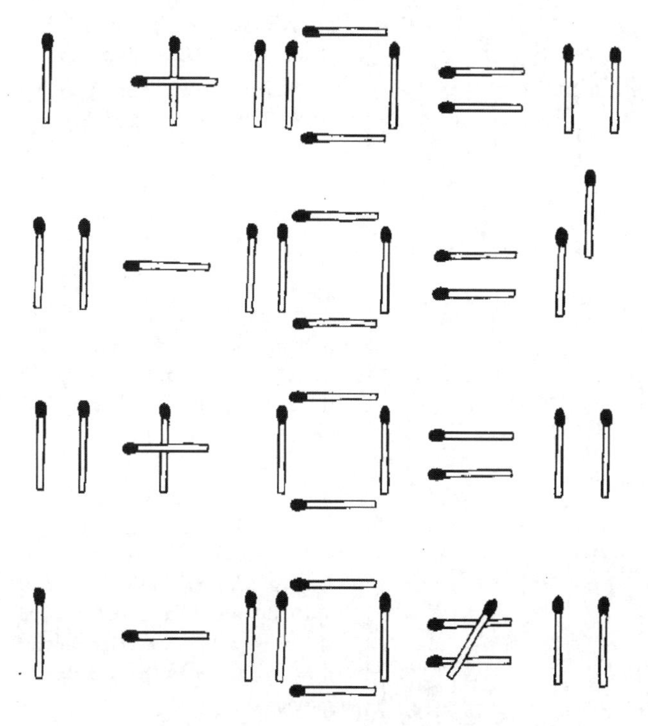

Start with the sticks as shown below:

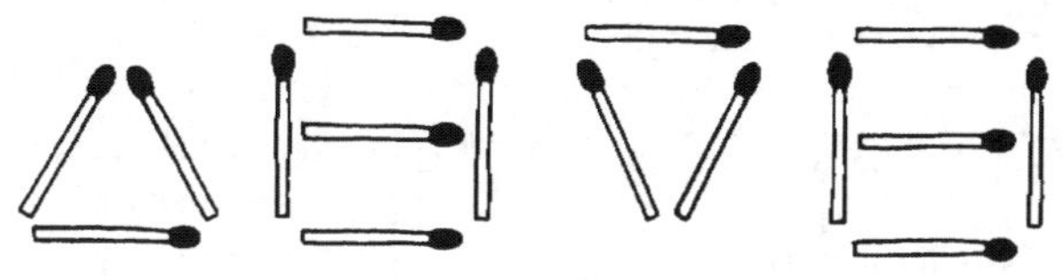

Take away four sticks to leave a word that spells what matches are made of. Most people try to make "WOOD"; however, the solution comes with recognizing a play on the word "match":

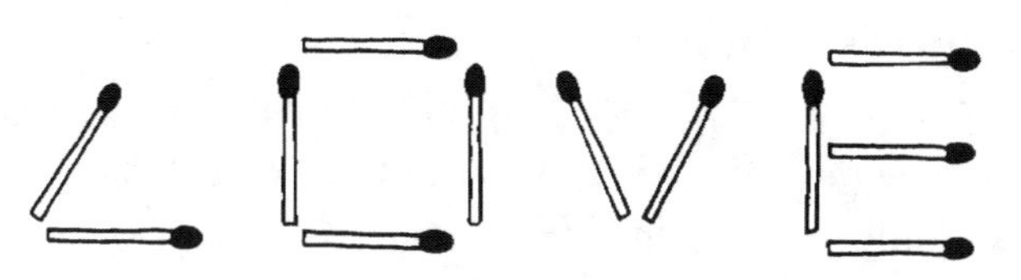

World's Smallest Crossword

世界上最小的纵横字谜

字谜是西方一个主要的文字游戏。特别是本节所提到的把字母重新排列后成为新字或短语。这要求对语言的熟练掌握。要知道 n 个不同字母的排列有 n！种。

Dr. Wordle: Now Mr. Hoi, here's a chance to win $ 20. This crossword puzzle is so simple that it only has six definitions. You have just three minutes to figure it out.

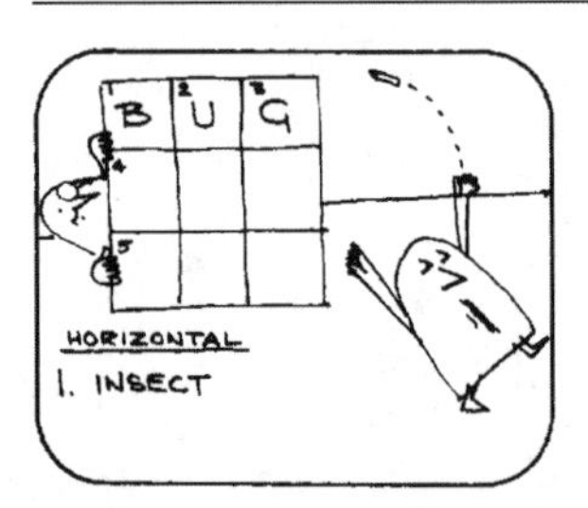

After three minutes, Mr. Hoi had done no more than guess the first row.

Mr. Hoi: Sorry, Dr. Wordle. I can't think of any other words that make sense.

Dr. Wordle: We're sorry too Mr. Hoi. You didn't see that all three horizontal words were the same. Remember, the same word can have different meanings.

Dr. Wordle: While we're waiting for our next guest here's another quickie for you home viewers. Can you take the seven letters in the two words "NEW" and "DOOR" and rearrange them to make one word?

Squares and Anagrams

Crossword puzzles are combinatorial problems that involve intersecting sequences of symbols. Now that computers can store in their memory all the words of a natural language, it is possible to write computer programs that solve crossword puzzles with great efficiency. Programs also can be written that will construct crossword puzzles.

Most crossword puzzles have patterns with "holes"—black cells that serve to separate words. An ancient and pure form of crossword, which has no holes at all (like our joke crossword puzzle), is known as a "word square". Here, for example, is a word square of order 4 (four-letter words):

K I N G

I D E A

N E X T

G A T E

Four words ("king", "idea", "next" and "gate") can be read both horizontally and vertically. If the horizontal and vertical words differ, it is called a "double word square":

O R A L

M A R E

E V E N

N E A T

The higher the square's order, the more difficult it is to compose word squares of either type. You might enjoy trying to compose some order-4 squares. If successful, try orders 5 and 6. Order-7 squares are extremely difficult. Although squares of orders 8, 9 and 10 have been constructed by word play experts, they almost always require the use of unusual, obscure words.

Dr. Wordle's "NEW DOOR" puzzle belongs to a puzzle category called "anagrams"—the rearranging of the letter of a word, phrase or sentence to make a new word, phrase or sentence. (For the solution, see the back of the book.) There are thousands of amusing anagrams in which the second permutation of letters has a meaning that is related in some appropriate way to the original permutation:

Lawyers: Sly ware.
Halitosis: Lois has it!
Punishment: Nine thumps
The Mona Lisa: No hat, a smile.
One hug: Enough?
The eyes: They see.
The nudist colony: No untidy clothes.

Maybe you can invent some better ones. It also is fun to rearrange the letters of your own name, or a friend's, or the name of a famous person to make an appropriate phrase or sentence. Here are two classic anagrams of William Shakespeare:

I ask me, has Will a peer?
We all make his praise.

Mary Belle Byram

玛丽·贝尔·拜伦

回文是最常见的文字游戏，汉语中也有。正读与反读完全一样，如“唧唧复唧唧”，此外还有许多回文诗。

Dr. Wordle's next guest is Mary Belle Byram. What is so unusual about her name?

Maybe the words on this sign will help you. The have the same property.

"Hat, Utah" and "Mary Belle Byram" are palindromes—sequences of letters that read the same backwards and forwards.

ADA

Can you compose other names of people that have palindromic symmetry? (It's not as easy as you might think!) Here are some examples:

Leona Noel
Nella Allen
Blake Dana de Kalb
Edna Lalande
Duane Rollo Renaud
N. A. Gahagan
N. Y. Llewellyn
R. J. Drakard, Jr.

Picture Puzzles

Dr. Wordle: Welcome to the show, Mary. Your first problem involves pictures. Each represents a familiar mathematical term.

Mary: I don't know what you mean, Dr. Wordle.

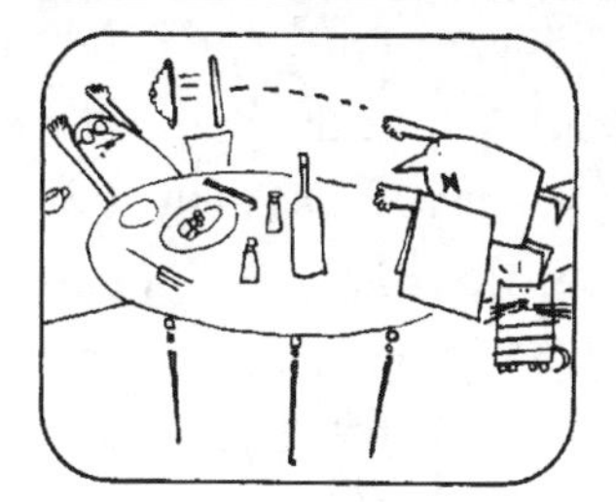

Dr. Wordle: Well here's an example. This picture stands for the geometrical constant "pi".

Mary: Oh, I see. I just have to guess what the picture stands for.

Dr. Wordle: Right you are. Now try to guess the others for $ 10 each. Here's the first.

Mary: I've got it. It's "polly gone", or "polygon".

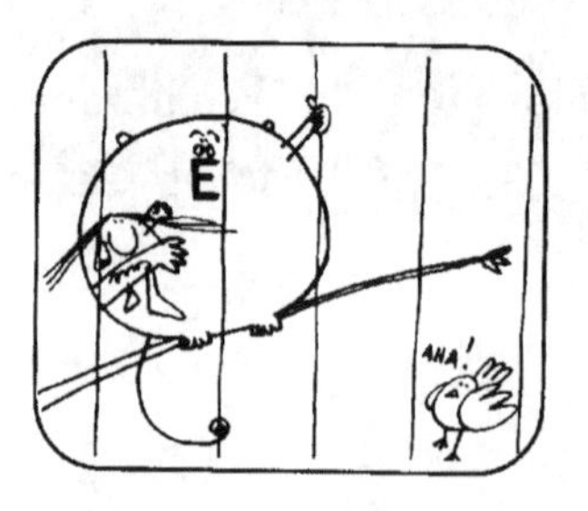

Dr. Wordle: Yes, it is. Now what about the second one?
Mary: Hmm. The lips are shaped like an "E". Is it "ellipse"?

Dr. Wordle: It sure is, Mary. See if you can get this last one.
Mary: Oh, that one's easy. It's a "radical".

Picture Words

用图表示词义十分形象，在广告中常用。

Pictures that represent words in some puzzling way are called *rebuses*. You might try your skill at inventing other rebuses for other familiar mathematical terms.

A cousin to the "rebus" is the writing of a word so that it somehow depicts the meaning of the word. The term "mathematics" has been used for mathematical words and phrases designed in this way. Figure 1 will give you the basic idea. Mathematics is a refreshing variant of the older rebus puzzle.

Drawing words so that they become symbolic pictures of their meaning is an important aspect of modern advertising, especially in advertising for motion pictures. Movie titles are frequently printed in such a way that the lettering symbolizes the title's meaning. (See Figure 2.) Artists often do the same thing with titles on book jackets. The use of symbols in traffic signs on streets and highways is another instance of how words and symbols can be combined to dramatize the word's meaning.

1

Examples of mathematics

TOPOLOGY

BIS ECT

GRAPH

SINEWAVE

DILATATION

LIMIT POINT

E^{x}PONENT

SUB$_{\text{SCRIPT}}$

TRANSLATION

PE RI OD IC

A D D

MULTIPL Y

RO T A T I O N

SECTION INTER

(MAT RIX)

2

British poster

Crazy Sentences

滑稽的句子

Dr. Wordle: Your next task, my dear, is to tell me what is so remarkable about each of the sentences I'm going to show you. You'll get $20 for each one you solve.

Dr. Wordle: Here's the first one. Read it carefully. And please stop tickling me.

Mary: I can't help it. You're so handsome that... that I want you to be tickled pink with me.

Dr. Wordle: Tickling me isn't going to get you $ 20.

Mary: Alright then. This sentence is a palindrome just like my name. It spells the same both ways.

Dr. Wordle: Very good, my love. How about this one?

Mary: Let't see. It's almost a palindrome but not quite.

Hmm. I've got it. It reads the same when you turn it upside down.

Dr. Wordle: Right again Mary, dear. Now for the last one.

Mary: I see the pattern. Each word has one more letter than the word before.

Dr. Wordle: Splendid. Here's another $ 20. What are you going to do with all that money?

Mary: I'm taking you to dinner tonight, sweetie, then to my apartment to show you my collection of dictionaries.

Dr. Wordle: Alright, Mary. That's a good idea. See you then. Now we have time for another quickie before our next guest comes on.

Dr. Wordle: What five-letter word does every Harvard graduate pronounce "wrong"?

回文至此才同数学游戏结合起来！最有趣的是回文型素数，如30 103、30 203。更一般的是回文素数对，即一个素数的反序也是素数，如13和31、17和71、113和311。更进一步还有一个素数和它的所有循环排列也都是素数，如11 939、19 391、93 911等。这些可以推广到更为一般的情形，例如，一个数和它的反序数加、减、乘、平方、立方之后是否有什么新花样？回文的确给我们带来一个好的数学游戏！

More Palindromes

Thousands of marvelous palindromes have been constructed in every major language. They are not hard to invent, and you may wish to try creating some yourself. Here are a few famous examples:

A man, a plan a canal—Panama!

Egad! A base tone denotes a bad age.

Was it a can on a cat I saw?

Live dirt up a side track carted is a putrid evil.

Ten animals I slam in a net.

In the classic palindrome, letters are the units. Palindromes can also be written in which words are the units. Here are two excellent specimens by J. A. Lindon, a British expert:

1. You can cage a swallow, can't you, but you can't swallow a cage, can you?

2. Girl bathing on Bikini, eyeing boy, finds boy eyeing bikini on bathing girl.

Poems have also been written by palindromists in which either the letter is the unit, or the word, or the line.

Palindromes are analogs of what mathematicians call bilateral symmetry. Humans and most animals are bilaterally symmetric. Many man-made objects also have bilateral symmetry: Chairs, for example, coffee cups, and thousands of other things. Any bilaterally symmetric figure, on the plane or in three dimensions, looks the same in a mirror. This is the analog of the palindrome's property that if you reverse the order of its symbols, the sequence is unchanged.

Numerals, like letters, also are symbols, and a palindromic number is simply a number that is the same when you read its digits in both directions. There is a famous unsolved number problem called the "palindrome conjecture". Take any number whatever, in decimal form, reverse it, and add the two numbers. Now repeat this procedure by reversing the sum and adding it to itself, and continue doing this until you get a palindrome. For example, 68 generates a palindrome in just three steps:

$$
\begin{array}{r}
68 \\
+\ 86 \\
\hline
154 \\
+\ 451 \\
\hline
605 \\
+\ 506 \\
\hline
1{,}111
\end{array}
$$

The palindrome conjecture is that no matter what number you start with, you will arrive at a palindrome after a finite number of steps.

No one yet knows whether the conjecture is true or false. It has been shown false for all numbers in binary notation, or any notation based on a power of 2. It has not yet been proved for numbers in any other notation.

The smallest decimal number that may be a counterexample to the conjecture is 196. Computers have carried it to hundreds of thousands of steps without obtaining a palindrome, but nobody has yet proved that it will never produce one.

Mathematicians have also investigated palindromic numbers that are also primes (numbers with no factors except 1 and themselves). It is believed that there are an infinite number of such palindromic primes, but this is not yet proved. It is also conjectured that there are an infinite number of palindromic prime pairs such as 30, 103 and 30, 203, in which all digits are the same except the middle digits, which are consecutive.

A palindromic prime must have an odd number of digits. Every palindromic number with an even number of digits is a multiple of 11, and therefore not a prime. Can you prove that such a number is always divisible by 11? (Hint: A number is divisible by 11 if the difference between the sum of all digits in even positions, and the sum of all digits in odd positions, is a multiple of 11.)

Square numbers are unusually rich in palindromes, such as 11 × 11 = 121. A square number is much more likely to be palindromic than an integer picked at random. The same is true of cubes. Moreover, a palindromic cube is almost certain to have a cube root that is also a palindrome (example: 11 × 11 × 11 = 1331). A computer search for palindromic fourth powers has failed to find a single one that has a fourth root that is not a palindrome. No one has yet found a fifth power that is a palindrome. It is conjectured that there is no palindromic number of the form x^k, where k is greater than 4.

The sentence "NOW NO SWIMS ON MON," is one of the longest ever discovered with "two-fold symmetry"—that is, it is unchanged by a 180-degree rotation. There are many examples of single words with the property, either printed or in longhand. Figure 3 shows a few.

3 Invertible signature and upside-down words

WH Hill

chumps

bunq

NOON

honey

The sentence, "I do not much enjoy dancing gorillas", is called a "snowball" sentence because successive words grow in size like a rolling snowball. Here are two even more remarkable specimens:

I do not know where family doctors acquired illegibly perplexing handwriting; nevertheless, extraordinary pharmaceutical intellectuality, counterbalancing indecipherability, transcendentalizes intercommunications incomprehensibleness.

I am not very happy acting pleased whenever prominent scientists overmagnify intellectual enlightenment, stoutheartedly outvociferating ultrareactionary retrogressionists, characteristically unsupernaturalizing transubstantiatively philosophicoreligious incompre hensiblenesses anthropomorphologically.

The answer to Dr. Wordle's last quickie is: The five-letter word that Harvard graduates pronounce wrong is the word "wrong". It is easy to think of variations. What word do Californians pronounce best? Of all seven-letter words, which is spelled easiest? And so on.

Nosmo King

可笑的名字

The next guest is Nosmo King, President of a cigarette company in Hackettstown, New Jersey. Do you see why Dr. Wordle was amused by his name?

一个小幽默!

If you shift the spacing between the first and last names, Nosmo King becomes "No Smoking". Isn't that something?

CHO PHO USE

Although this may seem trivial, it brings out the importance of the empty space as a symbol essential to the understanding of sentences. Spaces between words play a role analogous to such arithmetical symbols as parentheses, spacings, zero, and so on. The meaning of a mathematical expression is often completely altered by a minor change in the position of one parenthesis, just as "No Smoking" is drastically altered by shifting the position of the space between the two words.

Many words can be changed in meaning by putting a space inside them. "Nowhere", for instance, becomes "now here". Lewis Carroll wrote a short story about a man who thinks he has seen a sign that said "Romancement", when it actually said "Roman cement".

Here is an old puzzle sign said to have been attached to a post on a village street back in horse and buggy days:

TOTI EMU LESTO

Can you make sense of it by altering the spacing?

A perennially popular type of puzzle, along similar lines, is finding names concealed in a sentence. For example, a state and its capital are concealed within:

Can Eva dance outside, with cars on city streets?

The hidden words, shown above, are "Nevada" and "Carson City".

See if you can find the states and their capitals in the following sentences:

Al, ask Anne and June a useful question!

Ken, tuck your shirt in and be frank, forthright, and courageous.

Go north, Carol, in a car owned by Flora Leigh.

Are you afraid a hobo is entering your house?

This is where I connect, I cut the hart for dinner.

Mathematical terms are just as easily hidden. For instance, "A happy ram identifies a good farm," conceals a familiar name for a geometrical solid.

It is even possible to construct sentences in which a series of concealed words form a second sentence, and when those words are deleted, the letters that remain spell a third sentence! For example:

H*one* sh*all*owed fea*the*r a*corn*s w*is*e rest*rain*ed.

The italicized words spell: "*on all the corn is rain*." And if those words are deleted, the remaining letters spell: "*He showed fear as we rested.*"

This type of construction has its arithmetic analogs. For instance: 15 + 11 = 26. The underlined numbers form 5 + 1 = 6, and if these are deleted, 1 + 1 = 2 remains. Maybe you can invent more complicated examples.

Square Family

方卡片中的家谱

Dr. Wordle: Your first problem, Nosmo, for six boxes of fine Cuban cigars, has to do with this square card which contains the names of four people in a family.

Dr. Wordle: It's easy to put each name in a separate compartment by drawing three straight lines, but can you do it with just two straight lines?

Mr. King puffed silently on his cigar until his time was up.

Mr. King: It can't be done.

Dr. Wordle: Wrong, Mr. King. It's simple. That cigar smoke must be fogging your brain.

Straight and Equal

The aha! here is the realization that each name can be broken into two parts, and the parts combined in a different way to form the same four names.

一个数学趣题。但很特殊。

Many puzzles have been based on the problem of drawing straight lines in such a way that pictures of objects on the page are each put into a separate region. Figure 4 is typical. Can you draw 3 straight lines that will put each circle in a separate region? The solution follows the insight that the regions need not be rectangular, and that 3 lines can intersect in such a way that as many as 7 regions are created.

Interesting variations of this idea involve numbers instead of circles. The puzzle is to draw straight lines so that the sums of the numbers in each region are the same, or so the numbers in each region all share some other common property. Try your skill at this type of division on Figure 5. The task is to draw four straight lines so that the numbers in each of 11 regions have the same sum of 10. The solution to this task appears at the back of the book.

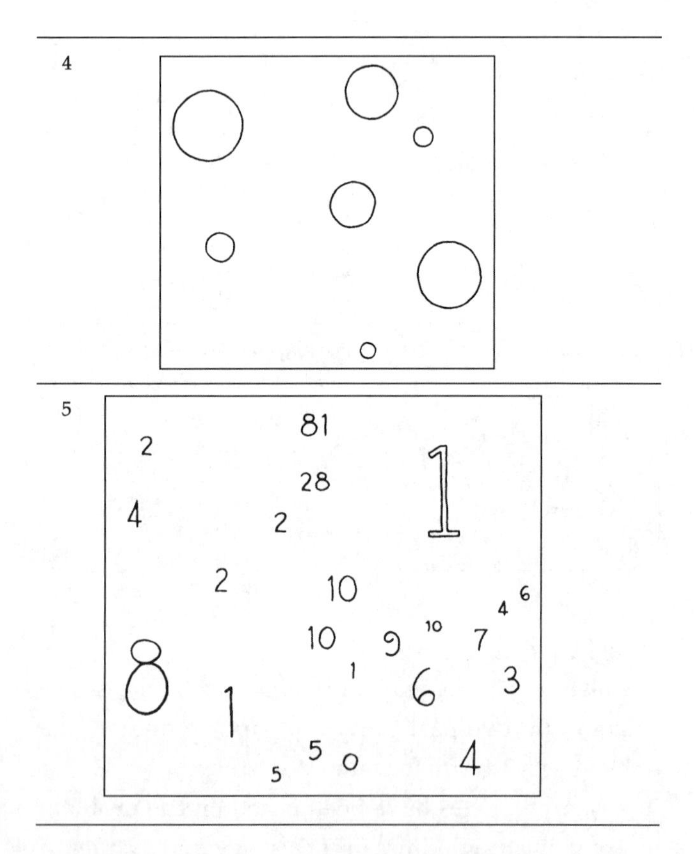

Tavern Tease

酒馆的招牌

Dr. Wordle: I'll give you another chance to win those six boxes of cigars. A tavern has this sign in its window.

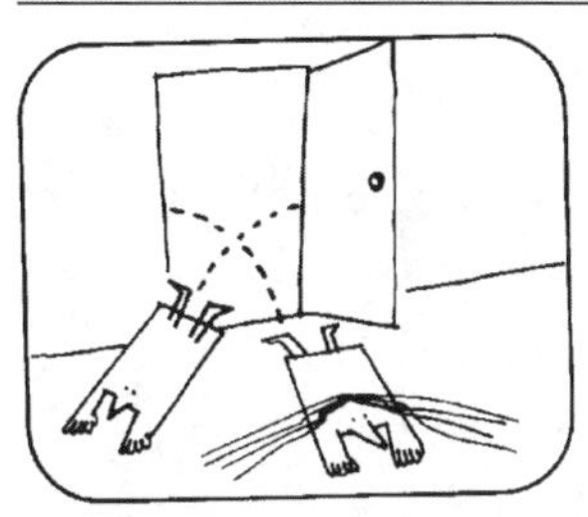

Dr. Wordle: But when kids under 18 go inside, they get tossed out for violating a law.

Dr. Wordle: The tavern owner says the sign painter forgot an exclamation mark and a question mark. Your task is to put in those two marks so that the sign says what the owner intended.

But Nosmo couldn't solve that one either. Dr. Wordle had to show him how to do it.

Marks and Signs

Puzzles in which a nonsense statement is made sensible by altering punctuation can be found in many old puzzle books. Here is a poem that seems to describe impossible things:

这是一些有趣的数学问题，可供初学数学的小学生做练习。

Though seldom from my yard I roam,
I saw some queer things here at home.
I saw wood floating in the air;
I saw a skylark, bigger than a bear;
I saw an elephant with arms and hands;
I saw a baby breaking iron bands;
I saw a blacksmith, weighing half a ton;
I saw a statue sing and laugh and run;
I saw a schoolboy nearly ten feet tall;
I saw an oak tree span Niagara fall;
I saw a rainbow, black and white and brown;
I saw a parasol walking through the town;
I saw a politician doing as he should;
I saw a good man—and I saw some wood.

The poem makes sense if the semicolons are shifted to the middle of each line:

"...I saw wood; floating in the air
I saw a skylark; bigger than a bear
I saw an elephant; ..."

and so on to the end.

Once more there are analogs with number puzzles. Consider the following false equation:

$$1+2+3+4+5+6+7+8+9=100$$

The task is to make this a correct equation by altering the "punctuation" on the left side. Only plus and minus signs may be used, but the spacing between the digits can be altered to make larger numbers. Here is the only solution with as few as three signs:

$$123-45-67+89=100$$

The solution with the largest number of plus and minus signs is:

$$1+2+3-4+5+6+78+9=100$$

There are just nine other solutions:

$123-45-67+89=100$

$123+4-5+67-89=100$

$123+45-67+8-9=100$

$123-4-5-6-7+8-9=100$

$12-3-4+5-6+7+89=100$

$12 + 3 + 4 + 5 - 6 - 7 + 89 = 100$
$1 + 23 - 4 + 5 + 6 + 78 - 9 = 100$
$1 + 2 + 34 - 5 + 67 - 8 + 9 = 100$
$12 + 3 - 4 + 5 + 67 + 8 + 9 = 100$
$1 + 23 - 4 + 56 + 7 + 8 + 9 = 100$
$1+ 2 + 3 - 4 + 5 + 6 + 78 + 9 = 100$

The same problem can be posed with the digits in descending order. This has 15 solutions if we exclude, as in the previous problem, the use of a minus sign in front of the first number:

$98 - 76 + 54 + 3 + 21 = 100$
$9 - 8 + 76 + 54 - 32 + 1 = 100$
$98 - 7 - 6 - 5 - 4 + 3 + 21 = 100$
$9 - 8 + 7 + 65 - 4 + 32 - 1 = 100$
$9 - 8 + 76 - 5 + 4 + 3 + 21 = 100$
$98 - 7 + 6 + 5 + 4 - 3 - 2 - 1 = 100$
$98 + 7 - 6 + 5 - 4 + 3 - 2 - 1 = 100$
$98 + 7 + 6 - 5 - 4 - 3 + 2 - 1 = 100$
$98 + 7 - 6 + 5 - 4 - 3 + 2 + 1 = 100$
$98 - 7 + 6 + 5 - 4 + 3 - 2 + 1 = 100$
$98 - 7 + 6 - 5 + 4 + 3 + 2 - 1 = 100$
$98 + 7 - 6 - 5 + 4 + 3 - 2 + 1 = 100$
$98 - 7 - 6 + 5 + 4 + 3 + 2 + 1 = 100$
$9 + 8 + 76 + 5 + 4 - 3 + 2 - 1 = 100$
$9 + 8 + 76 + 5 - 4 + 3 + 2 + 1 = 100$

If a minus sign is allowed in front of the first number, there are three more answers for the descending series and one more for the ascending:

$-9 + 8 + 76 + 5 - 4 + 3 + 21 = 100$
$-9 + 8 + 7 + 65 - 4 + 32 + 1 = 100$
$-9 - 8 + 76 - 5 + 43 + 2 + 1 = 100$
$-1 + 2 - 3 + 4 + 5 + 6 + 78 + 9 = 100$

The "punctuation" need not, of course, be limited to plus and minus signs, nor need the sum on the right be 100. For example, the sum could be the last two digits of the current year, or any other number you like.

See if you can place just one pair of parentheses in the following equation to make it correct:

$1 - 2 - 3 + 4 - 5 + 6 = 9$

The answer is at the end of the book.

Cryptic Symbols
隐蔽的符号

文字或字母的游戏从直观看两种性质最突出：一是对称性质，一是拓扑性质。

Dr. Wordle: Now, Mr. King, we're going to show you three pictures of strange symbols. Each conceals a word. Solve any of them and you get the cigars. Here's the first one. Do you see the solution?

Mr. King: No. It sure beats me. What is it?

Dr. Wordle: It's your first name, Nosmo. We made the symbols by reflecting your name below a line the same way a shoreline is reflected in a lake.

Dr. Wordle: Alright, maybe you'll get the next one.

Mr. King just shook his head as Dr. Wordle explained.

Dr. Wordle: This time the symbols were obtained by reflecting each letter over a vertical axis of symmetry. See how simple it is?

Mr. King: It's not so simple for me.

Dr. Wordle: Well, here's the last one. You still have a chance.

Mr. King was still unable to answer and Dr. Wordle had to add black lines above and below the symbols to bring out the word "SMOKE".

Symmetrical Fun

In the first set of strange symbols, each letter is reflected by a horizontal axis of symmetry. Note that some letters in NOSMO KING are unchanged by the reflection. These are the letters "O", "K" and "I", which have a horizontal axis of symmetry.

对称性质是图案中常用的，这里对称性是广义的，如序列的对称。

In the second set of symbols, each of which is reflected about a vertical axis of symmetry, certain letters also are unchanged. These are the letters "O", "M" and "I", which have vertical axes of symmetry. Because "O" and "I" have both kinds of axes, these letters are unchanged when reflected by a mirror above or below, or on either side. You might be interested in analyzing all the letters of the alphabet, in both capital and lower case form, to see what kind of symmetry each has.

Can you construct a word which is unchanged by a mirror held above it? Yes, the word "CHOICE" is one of hundreds of such words. Are there words which when printed vertically are unchanged by a mirror on the side? Yes, the word "TOMATO" is one of hundreds of such words.

Any plane figure with at least one axis of symmetry looks the same in a mirror, although one image may have to be rotated to give the same orientation to the object and its image. Any solid figure with a plane of symmetry also looks the same in a mirror. The reason we look the same when we see our mirror reflection is that a plane of symmetry bisects our body from head to toes.

Many interesting variations of our two mirror puzzles can be devised. What, for example, do we have here?

This one is even harder to recognize:

The letters of SMOKE are disguised in an entirely different way. The mind tends to see the black areas as odd-shaped figures rather than see the white spaces in between as the shapes that are the letters. It is like looking at the negative of a photograph. When there are no horizontal boundaries shown above and below the word, it is difficult to see the word. Try printing other words in a similar fashion.

Gold Tuitt

镀金的模型飞机

Dr. Wordle: Sorry, you didn't win the cigars, Mr. King. But you were such a good sport that I'm going to give you this gold plated "tuitt".

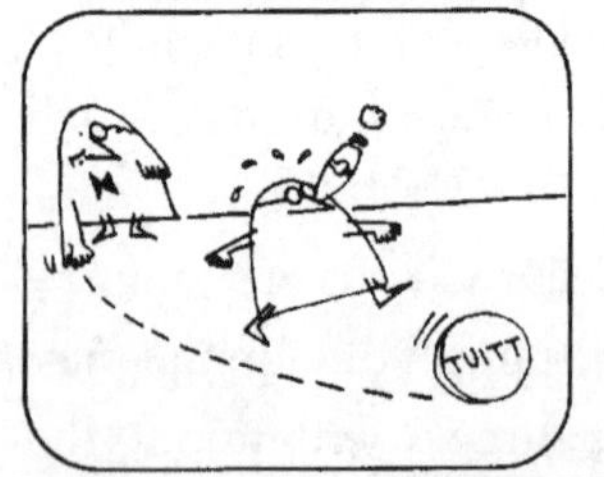

Mr. King: Thank you, Dr. Wordle. But what in the world is a "tuitt"?

Dr. Wordle: Isn't there something you've always wanted to do if you could only get around to it?

Mr. King: Yes. I've always wanted to learn how to fly a plane.

Dr. Wordle: Well, now you've got a round tuitt! Good luck Mr. King. And thanks for being with us.

Dr. Wordle: While our next guest is in make-up, I've got another quickie for my audience. This is the Christmas card I sent to all my friends last year. Can you find its secret message?

Flo Stuvy

费罗·斯特菲

The last guest on the show is Miss Flo Stuvy. Why do you think she was picked as a contestant?

The letters in her name are in alphabetical order. Names like this are not easy to find. See how many you can search out in the phone book.

Agony

Names in alphabetical order are not easy to find. "Betty" is a common example. One of the longest that uses actual first and second names is Abbe F. Gillott. Can you find any words of more than four letters that are alphabetized? "Billowy" is one of the longest. Short words, like "Dirt", are easy to find, but longer words are much harder to discover.

From time to time poems have been written in which the initial letters of the words run from A to Z. One of the best is John Updikes's poem "Capacity" in his book *The Carpentered Hen and Other Tame Creatures.*

Curious Sequences

奇妙的字母序列

拓扑是几何图形最粗的性质。

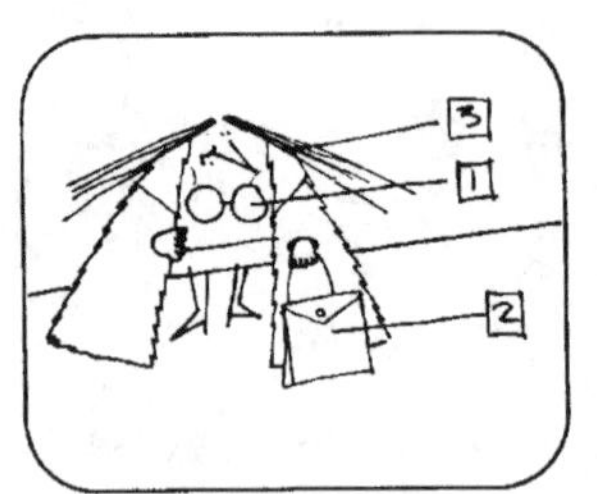

Dr. Wordle: Are your brains groovy, Miss Stuvy? We're going to show you three letter sequence problems. Solve one and you get a bathing suit. Solve two and you get a purse to go with it. Solve all three and you get a mink coat as well.

Dr. Wordle: Here's the first one. Notice that some of the letters are red and the rest blue. What rule did the artist follow when he partitioned the alphabet into these two colors?

Miss Stuvy studied the letters for almost a minute before replying.

Miss Stuvy: Eureka! Every red letter has at least one curved line while the blue letters are made up entirely of straight lines.

Dr. Wordle: You've won the bathing suit, Miss Stuvy. Now let's try for the purse. What is the rule for the partitioning of *this* alphabet into red and green letters?

Miss Stuvy: Let's see. It's not curves. It's not holes. It's not letters that rhyme. Hmm. Aha! I see the rule. The red letters are all topologically equivalent. They're all like a straight line.

Dr. Wordle: Good show, Flo. Now for the mink coat. See if you can cross out just six letters in the following sequence and leave an ordered set of letters that spells the name of a famous poet.

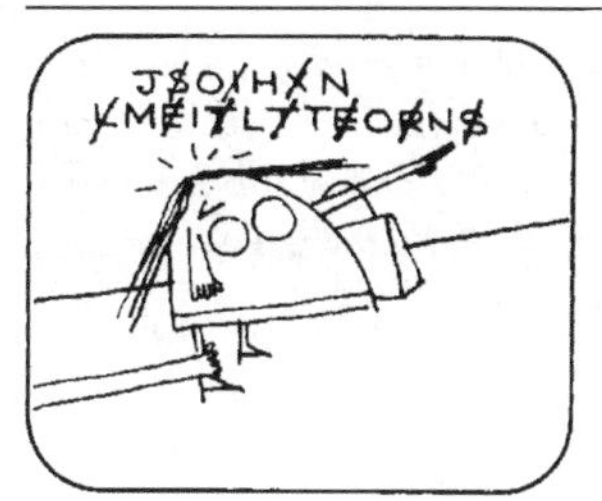

Miss Stuvy thought for quite a while before she got the right insight. Then she solved the problem neatly by crossing out S-I-X-L-E-T-T-E-R-S, "six letters."

She was so happy with her gifts that she gave Dr. Wordle a big hug and kiss.

Topology of the Alphabet

The first problem hinges on the geometrical difference between straight lines and curves. The second is based on the topological difference between a simple open curve and one that closes or branches.

> 最简单的拓扑学——一维的拓扑不同之处在于封闭○还是开放——。

Think of a capital letter as made of an elastic material that can be stretched or contracted, or even picked up from the plane and replaced in a different spot. Two letters are topologically equivalent if one can be changed to the other by such a deformation process. You are not permitted, of course, to cut a letter apart, or attach parts of it to itself. An interesting exercise is to classify all the capital letters into their topologically equivalent classes.

For example, E, F, Y, T and J are topologically the same, but not the same as K and X which belong to a different class. Similar classifications can be made of lower case letters, and also numerals, but you must watch out for variations in the way letters are printed.

Parting Words

最后的话

这些是纯文字游戏，对你的英语能力是个挑战!

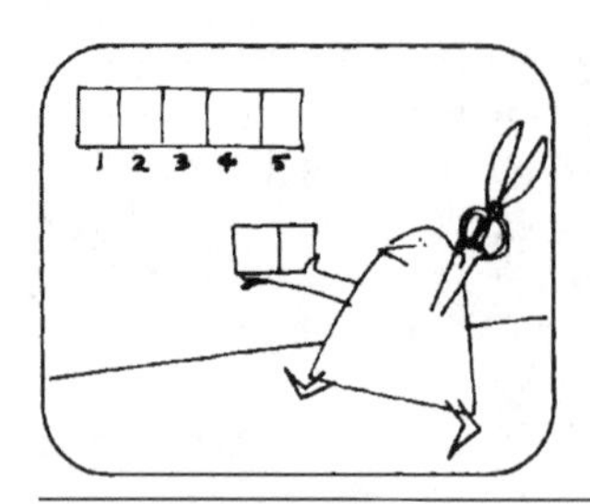

Dr. Wordle: And now viewers, I have three more quickies that I will leave with you. First, what five-letter word becomes shorter when you add two letters to it?

Second, what four letter word ends with "ENY"?

And third, can you think of a nine-letter word that has only one vowel?

Dr. Wordle: That's all for tonight, "wordlers". You were a great audience. See you next week. Same time. Same station.

Last Words

The quickies are answered as follows:

1. "Short" becomes "shorter" when two letters are added.

2. The four-letter word ending in "eny" is "deny".

3. The nine-letter word with only one vowel actually appears in the picture in singular form. The word is "strengths".

Here are some more word quickies of a similar sort:

1. Name a state that begins with 10 and is not Tennessee.

2. Name a state that ends with 10.

3. Rearrange the letters of "CHESTY" to spell another English word of six letters.

4. Read this couplet so that it rhymes correctly:

There was an old lady and she
Was deaf as a post.

5. Which word doesn't belong on this list?

Uncle
Cousin
Mother
Sister
Father
Aunt

6. What do these pairs of letters represent?

ST ND RD TH

7. Read these two sentences:

$\frac{\text{UALLS}}{\text{NOW}}$ WETHER

The answers all appear at the back of the book.

Answers to Posed Problems
答　案

Chapter 2—Geometry aha!

Devilish Divisions: Dissection Theory

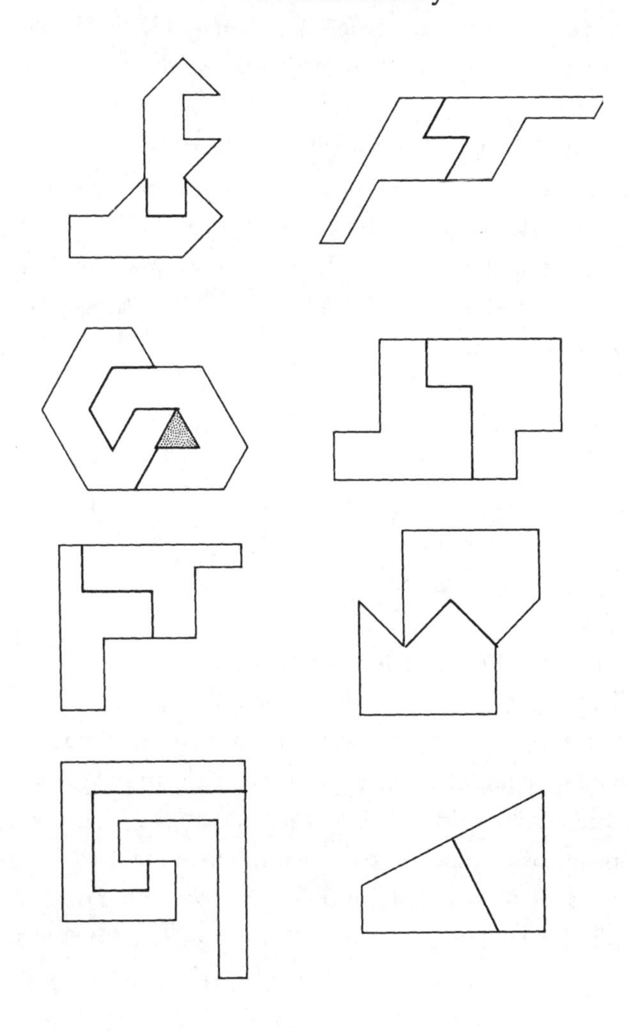

Chapter 3—Number aha!

Broken Records: Half Wholes

42

Eyes and Legs: Bipeds and Quadrupeds

The insight that leads to the solution is the realization that some animals have no feet—namely snakes! Once you have this aha! insight, it is not hard to find the only solution: four 4-footed beasts, two 2-footed animals, and five snakes.

The Big Bump: Thinking Backward

Did you answer this by saying that it took one-third as long as in the original problem—namely 12/3=4 hours? If so, you are wrong. The answer is exactly the same as before!

In the original version, the first spore becomes three at the end of the first hour, which is precisely the way the variation *begins*. Therefore, if it took 12 hours for the container to become filled in the original problem, in the variation it takes one less hour, or 11 hours. The container is filled at 11 o'clock.

Uncle Henry's Clock: Setting the Clock

If a clock takes 5 seconds to strike 6, there is a one-second interval between strikes. Therefore, it will take 11 seconds to strike 12.

Uncle Henry slept 40 minutes.

Spirits of 1776: Modulo Arithmetic

The insight, like the one that solved two of our previous problems, is to time reverse the procedure. Hold the king of spades face down in one hand. Pick up the queen, which has a value of 12, put it beneath the king, then one at a time move 12 cards from *bottom to top*. Pick up the jack (value 11), put it on the bottom of the packet, and transfer 11 cards, one at a time, from bottom to top. Continue in this manner, which is merely a time reversal of the Josephus count, and you will end with a packet of 13 cards in correct order.

The Josephus count need not be limited to consecutive numbers. The procedure just described arranges a packet for a Josephus count in which the numbers are entirely arbitrary; that is, they may be any numbers whatever, and in any desired order!

This can be strikingly demonstrated by the following card trick using the same packet of 13 spades. Instead of counting numbers, however, we spell the name of each card by moving a card from top to bottom for each letter. Begin with the cards in the following order from top down: Q, 4, A, 8, K, 2, 7, 5, 10, J, 3, 6, 9. Spell A-C-E, moving cards one at a time to the bottom. The card for E is turned face up. It is the ace. Put the ace aside, then spell T-W-O, and so on for the other cards until all thirteen have been spelled.

The initial arrangement of the cards was obtained by the same time-reversed procedure described above. Indeed, one can arrange an entire deck of 52 cards in this way so that all 52 can be spelled, using the full name for each card, such as A-C-E-O-F-S-P-A-D-E-S, and taking them in, say, spades, hearts, clubs, diamonds order.

The Josephus count procedure is so general that it obviously works with the spelling of any name whatever. For example, you can prepare a deck of file cards bearing any kinds of pictures you like — animals, zodiac signs, faces of famous people, and so on. The time-reversed procedure enables you to arrange the cards so you can go through the deck, spelling the name of each picture, and always dealing the corresponding card at the end of each spelling.

Chapter 4—Logic aha!

Six Sneaky Riddles: Sneaky Answers

1. The man had salted his soup before he noticed the fly.

2. The water will never reach the porthole because the ship rises with the tide.

3. The Hudson River was frozen near the shore when the Reverend Sol Loony walked on it.

4. One train went through the tunnel an hour after the other train had gone through it.

5. The convict was near the end of a long bridge. He had to run toward the approaching police car to get off the bridge before the car reached him.

6. 1,977 dollar bills are worth $1,977 and 1,976 bills are worth only $1,976.

The Big Holdup: Missing Evidence

If you are familiar with casette tape recorders, you will know that had Jones stopped the recording when Smith entered the room, the tape would not have been rewound. The real killer must have listened to the recording several times to make sure it sounded authentic, and then made the fatal mistake of leaving the tape rewound.

Dr. Ach's Tests: Dr. Ach's Solutions

1. Bend the match in the middle before you drop it.

2. By pouring sand slowly into the hole, the baby bird is raised to the top.

3. Form a small loop in the string, and tie a knot at its base. Cut the loop.

4. A 20-centimeter piece, cut from the pole, has a rectangular cross section of 20 centimeters by 50 millimeters (or 5 centimeters), therefore, it will fit the hole snugly.

5. With the ruler, measure the bottle's inside diameter and the height of the liquid. The liquid forms a cylinder, so its volume is easily calculated. Turn the bottle upside down. The air space now forms another cylinder of smaller height, also easily measured and its volume determined. Addition of the volume of air space to the volume of liquid gives the bottle's total capacity. The percentage of liquid is now easily calculated. Since the two cylindrical volumes have the same diameter, actually only their heights need be measured to get the percentage.

Barbershop Banter: Surprising Solutions

1. He suggested that each driver drive another man's car. The billionaire had offered the prize to the man whose *car* came in last, not the driver himself.

2. Hold the burning match under a glass of water.

3. The theater was a drive-in.

4. He goes into another room, gets on his hands and knees, and "crawls in" to the bottle.

5. The score of any ball game, before it starts, is nothing to nothing.

6. The man was a minister.

7. The myna bird was deaf.

8. Push the cork into the bottle.

Murder At Sun Valley: The One Way Ticket

1. The surgeon was the boy's mother.

2. The Frenchman kissed his own hand, then bashed the Nazi officer in the face.

Foul Play at the Fountain: Mirror Vision

1. The slave turned the box upside down, then slid the lid back just enough to allow a few diamonds to fall out.

2. The lady was on foot, not in a car.

Chapter 6—Word aha!

World's Smallest Crossword: Squares and Anagrams

The answer to Dr. Wordle's quickie is: The letters of "NEW DOOR" can be rearranged to spell "ONE WORD".

Square Family: Straight and Equal

The 11 regions formed by drawing four straight lines on top of Figure 5 of Chapter 6, Word aha!, are shown in the following figure:

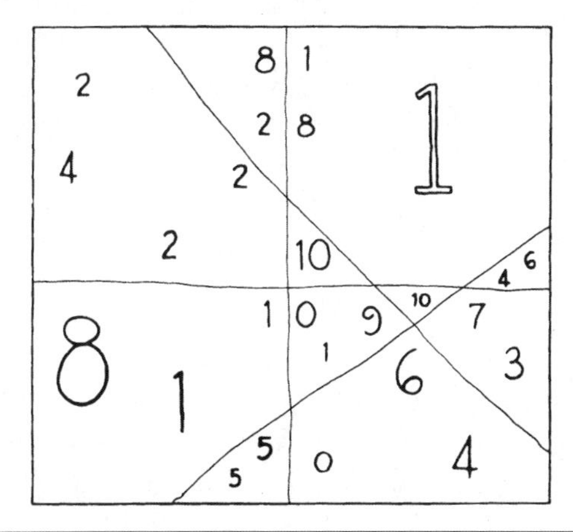

Tavern Tease: Marks and Signs

1- (2-3+4-5) +6=9

Parting Words: Last Words

1. Iowa.

2. Ohio.

3. Scythe.

4. There was an old lady and she was deaf as a P. O. S. T.

5. Cousin—it is the only word that does not indicate a person's sex.

6. The letter pairs are the endings of "first", "second", "third" and "fourth."

7. All between us is over now.

A bad spell of weather.